SIMPLY

EMERGING TECHNOLOGY

FACTS MADE FAST

DK LONDON

Senior Editor Alison Sturgeon
Senior Designer Mark Cavanagh
Editors Claire Cross, Rob Dimery,
Dorothy Stannard
US Senior Editor Jennette ElNaggar
Designers Vanessa Hamilton,
Mark Lloyd, Lee Riches
Managing Editor Gareth Jones
Senior Managing Art Editor Lee Griffiths
Production Editor Robert Dunn
Senior Production Controller Rachel Ng
Jacket Design Development Manager
Sophia M.T.T.
Jacket Designer Akiko Kato
Associate Publishing Director Liz Wheeler
Art Director Karen Self
Publishing Director Jonathan Metcalf

First American Edition, 2024
Published in the United States by DK Publishing
1745 Broadway, 20th Floor, New York, NY 10019

Copyright © 2024 Dorling Kindersley Limited
DK, a Division of Penguin Random House LLC
24 25 26 27 28 10 9 8 7 6 5 4 3 2 1
001–336902–Jun/2024

A catalog record for this book is available
from the Library of Congress.
ISBN 978-0-7440-9198-4

Printed and bound in China
www.dk.com

MIX
Paper | Supporting
responsible forestry
FSC™ C018179

This book was made with Forest
Stewardship Council™ certified
paper—one small step in DK's
commitment to a sustainable future.
Learn more at
www.dk.com/uk/information/sustainability

AUTHOR AND CONSULTANT

Hilary Lamb is an award-winning science
and technology journalist, editor, and
author. She studied physics at the
University of Bristol and science
communication at Imperial College,
London, before spending five years as a
staff magazine reporter. She has worked
on previous DK titles, including *How
Technology Works*, *The Physics Book*,
Simply Quantum Physics, and *Simply
Artificial Intelligence.*

CONTRIBUTOR

Bea Perks is a medical writer, science
writer, and journalist based in Cambridge,
UK. With a PhD in clinical pharmacology,
she has edited journals, written for
pharmaceutical companies, and
contributed to popular science
magazines and websites. She has
written for several DK titles, including
*Knowledge Encyclopedia Science!, Super
Science,* and *Timelines of Science.*

CONTENTS

BIOTECHNOLOGY

ROBOTICS

COMMUNICATIONS AND **MEDIA**

ENERGY

THE BUILT ENVIRONMENT

EMERGING AND FUTURE TECHNOLOGIES

From the invention of the wheel to the arrival of personal computers, technology has shaped the world we live in. Emerging technologies are broadly defined as those yet to enjoy widespread use. Some are limited to research environments; others are fully developed but vying for commercial viability. Introducing a new technology into the commercial arena can be a huge challenge—unless a technology can be successfully scaled up, its potential will not be fulfilled.

Many emerging technologies are likely to have a significant social, environmental, or economic effect. A great deal of work is being carried out in researching, developing, and implementing these technologies with a view to improving the world—whether through innovative treatments to help patients with rare medical conditions or genetically engineering crops to combat hunger on a global scale.

Much effort and investment is devoted to technologies that may help reduce greenhouse gas emissions and avert the worst effects of climate change. These technologies span clean energy infrastructure, alternatives to fossil fuel-powered transportation, and tools for optimizing industrial processes, to name a few.

New technologies are also emerging as a result of the increasing connectivity of the world. Tens of billions of digital devices collect and analyze data around the world every day. This information can be used to manage all sorts of systems efficiently, from helping individual cars navigate traffic to optimizing services across entire cities—with AI directing decision-making in real time. In this increasingly automated world, dull, dirty, or precarious work is being taken over by machines.

This moment represents an exciting and dynamic period of technological innovation. Used wisely, these emerging technologies could resolve many of the world's most urgent challenges.

MATERI

A L S

Materials engineering involves examining and adapting existing materials, and inventing new materials, to help solve problems. Today, there is great interest in developing materials that are more sustainable (compostable plastics and efficient electronic components); "intelligent" (self-healing substances and those that adapt to their environment); and useful in medical applications (biocompatible devices and 3D-printed organs). Some materials are inspired by nature, while others have properties never seen before, such as the ability to bend light around an object. Novel materials can become truly practical, however, only when they are affordable to make and use.

TOO SMALL TO BE SEEN

Nanomaterials are a diverse group of materials typified by their very small size. They have dimensions of less than 100 nanometres (nm); 1 nm is one-billionth of a meter. This is known as the nanoscopic scale, or nanoscale. Nanomaterials exist in nature but can also be engineered with certain desirable qualities, such as extreme strength or electrical conductivity. These properties give them potential applications across countless industries—for example, nanomedicine (see p.32) is an experimental field that uses technology on the nanoscale to help diagnose and treat conditions.

Four categories
Nanomaterials can be classified into four broad types according to how many dimensions are measured in the nanoscale.

0D

0D nanomaterials
Fullerene-C60 (or "buckyballs") and nanoclusters are examples of 0D nanomaterials—those with dimensions of less than 100 nm.

1D

1D nanomaterials
Long and thin, 1D nanomaterials, which include nanowires and carbon nanotubes, have a single dimension above 100 nm.

2D

2D nanomaterials
With two dimensions above 100 nm, all 2D nanomaterials consist of very thin layers. They include graphene, a nanosheet.

3D

3D nanomaterials
Comprising nanoparticles and nanolayers, 3D nanomaterials are "bulk" materials that measure above 100 nm in every dimension.

> ## "Nanotechnology is manufacturing with atoms."
> William Powell, lead nanotechnologist at the Goddard Flight Center

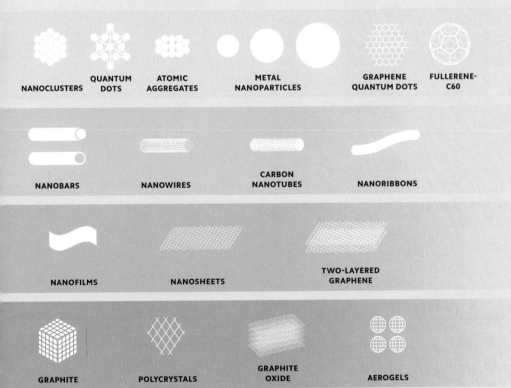

NANOCLUSTERS   **QUANTUM DOTS**   **ATOMIC AGGREGATES**   **METAL NANOPARTICLES**   **GRAPHENE QUANTUM DOTS**   **FULLERENE-C60**

NANOBARS   **NANOWIRES**   **CARBON NANOTUBES**   **NANORIBBONS**

NANOFILMS   **NANOSHEETS**   **TWO-LAYERED GRAPHENE**

GRAPHITE   **POLYCRYSTALS**   **GRAPHITE OXIDE**   **AEROGELS**

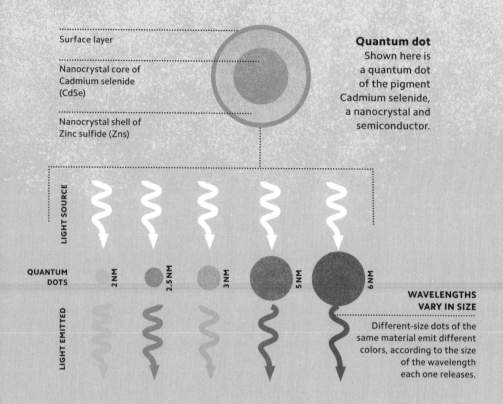

Surface layer

Nanocrystal core of
Cadmium selenide
(CdSe)

Nanocrystal shell of
Zinc sulfide (Zns)

Quantum dot
Shown here is
a quantum dot
of the pigment
Cadmium selenide,
a nanocrystal and
semiconductor.

LIGHT SOURCE

QUANTUM
DOTS

2 NM   2.5 NM   3 NM   5 NM   6 NM

LIGHT EMITTED

**WAVELENGTHS
VARY IN SIZE**

Different-size dots of the
same material emit different
colors, according to the size
of the wavelength
each one releases.

COLOR-EMITTING CRYSTALS

Quantum dots are nanoscale crystals with unique optical and electronic properties. Thanks to their very high surface-area-to-volume ratio, they are photoluminescent—they absorb and release light. When quantum dots are illuminated by ultraviolet light, they produce different frequencies of visible light, according to the size of the crystal. For instance, larger quantum dots emit low-frequency colors such as red and orange, while smaller ones emit high-frequency colors, including blue and purple. This could make them useful for new types of LEDs, lasers, and medical imaging devices, among other applications.

ENGINEERING BEYOND NATURE

Metamaterials are composite materials designed to have properties that have not yet been seen in nature. They are engineered from different substances—such as metals, plastics, and ceramics—which are arranged in repeating structures at scales smaller than the wavelengths of incoming light. This framework gives metamaterials extraordinary optical properties. For example, they could potentially be used to build an "invisibility cloak" that guides light around a cloaked object, hiding it from view. It is now also possible to manipulate sound waves, by the use of specially designed acoustic metamaterials.

WITHOUT INVISIBILITY CLOAK

MICROWAVE SOURCE

SOLID OBJECT

Bouncing back
Microwaves usually bounce back from an object, as does visible light (which has a higher frequency than microwaves). This allows us to detect the object.

WITH INVISIBILITY CLOAK

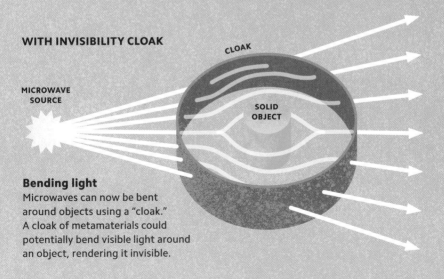

MICROWAVE SOURCE

CLOAK

SOLID OBJECT

Bending light
Microwaves can now be bent around objects using a "cloak." A cloak of metamaterials could potentially bend visible light around an object, rendering it invisible.

DAMAGE LIMITATION

Self-healing materials have the lifelike ability to repair themselves, without human intervention, after sustaining damage such as cracks or cuts. There are many ways in which they do this—some are triggered by an external stimulus, such as light or heat, while others need no stimulus other than the damage itself. Polymers are the most common kind of auto-restorative substance, although these also include metals, ceramics, and various concretes. Self-healing materials have the potential to last far longer than conventional materials and have many uses, including the construction of hardier roads, buildings, and satellites.

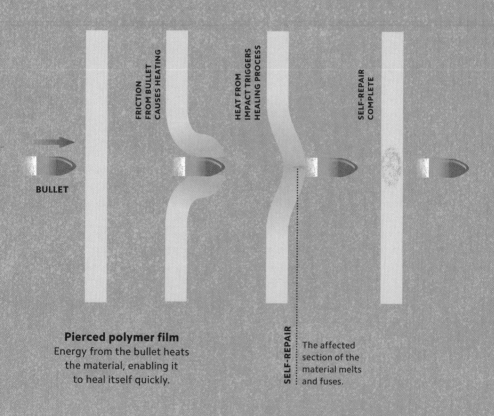

BULLET

FRICTION FROM BULLET CAUSES HEATING

HEAT FROM IMPACT TRIGGERS HEALING PROCESS

SELF-REPAIR COMPLETE

Pierced polymer film
Energy from the bullet heats the material, enabling it to heal itself quickly.

SELF-REPAIR
The affected section of the material melts and fuses.

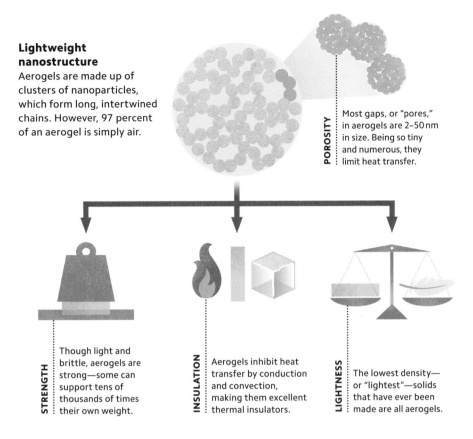

Lightweight nanostructure

Aerogels are made up of clusters of nanoparticles, which form long, intertwined chains. However, 97 percent of an aerogel is simply air.

POROSITY

Most gaps, or "pores," in aerogels are 2–50 nm in size. Being so tiny and numerous, they limit heat transfer.

STRENGTH

Though light and brittle, aerogels are strong—some can support tens of thousands of times their own weight.

INSULATION

Aerogels inhibit heat transfer by conduction and convection, making them excellent thermal insulators.

LIGHTNESS

The lowest density—or "lightest"—solids that have ever been made are all aerogels.

CLOUD-LIGHT, STEEL-STRONG

Aerogels are a class of ultralight material with properties that belie their almost cloudlike appearance. Although they are derived from gels, the liquid is replaced with a gas, such as air—hence the name—while still retaining the gel's original structure. The result is a strong solid with extremely low density, low thermal conductivity, and other advantageous qualities, depending on the type of aerogel. They have a range of applications, including thermal insulation (for example, in spacesuits and on NASA's Mars rovers), particle detectors, drug delivery systems, and sports equipment.

DRY ADHESIVES
The unique structures on geckos' toes, which allow them to run on almost any surface, are influencing new adhesives.

SUPER-STRONG MATERIALS
Spider silk is the toughest fiber in nature. Scientists are striving to develop artificial versions.

FASTER SWIMSUITS
Swimsuits that mimic shark skin's overlapping, grooved scales enable faster swimming.

INSPIRED BY NATURE

With the benefit of millions of years of evolution, nature has found ingenious solutions to all sorts of problems. Technological inspiration taken from the natural world is known as biomimetics and covers many fields, including robotics and materials science. Biomimetic materials have been around for many decades (Velcro, inspired by burr hooks, was invented in the 1940s), and new materials with unusual and useful properties are being engineered every year. They range from water-harvesting materials inspired by cacti and beetles to swimsuit fabric modeled on shark's skin.

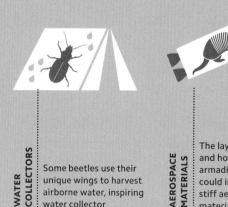

WATER COLLECTORS
Some beetles use their unique wings to harvest airborne water, inspiring water collector

AEROSPACE MATERIALS
The layers of bone and horn in armadillo armor could inform new stiff aerospace materials.

ADAPTIVE CAMOUFLAGE
Scientists hope to replicate the ability of cephalopods to control the appearance of their skin.

SHAPE MEMORY

COLOR-SHIFTING

SELF-REPAIRING

SELF-CHARGING

THERMAL CONTROL

SELF-CLEANING

Responsive wearables
Advances in fields such as
flexible electronics (see p.18)
mean that textiles can be
designed with different abilities.

COMPUTERS IN OUR CLOTHES

Smart textiles incorporate enhanced properties, such as the ability to
sense and respond to their environment or wearer. This allows for
clothing that is more comfortable, protective, and useful, or which
may even function as a wearable device if fitted with electronic
components. Such textiles may include fabric types that respond to
light, temperature, and sound. Some are able to disinfect themselves,
"remember" their shape, or gather data—such as health and fitness
information—from the wearer's body in real time.

FOLDABLE DEVICES

Electronic circuits are typically rigid, but by mounting the active components on flexible substrates, such as film, foil, or fabric, they can become pliable. The first flexible solar cells were designed in the 1960s. Since then, technological advances (such as in the field of printed circuits) have led to a huge range of devices that are capable of bending, rolling, twisting, stretching, and folding. For example, users can wear thin, flexible circuits equipped with integrated sensors on their skin for health monitoring.

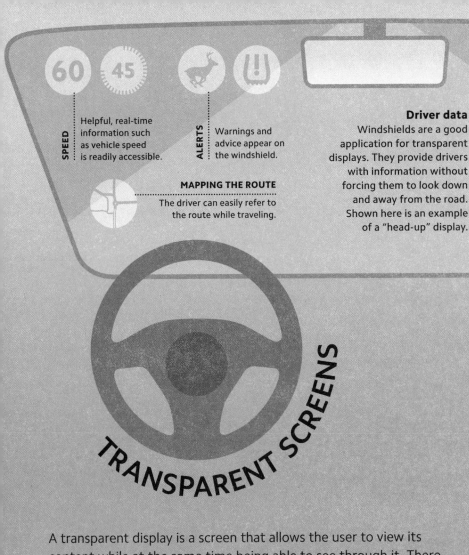

SPEED

60 45

Helpful, real-time information such as vehicle speed is readily accessible.

ALERTS

Warnings and advice appear on the windshield.

MAPPING THE ROUTE

The driver can easily refer to the route while traveling.

Driver data

Windshields are a good application for transparent displays. They provide drivers with information without forcing them to look down and away from the road. Shown here is an example of a "head-up" display.

TRANSPARENT SCREENS

A transparent display is a screen that allows the user to view its content while at the same time being able to see through it. There are two main types—emissive displays, which produce images by generating light, and absorptive displays, which do so by blocking light. Other types are currently under development, including one that will work on any transparent surface. See-through displays are useful in augmented reality systems, layering digital content onto the user's view of the world, such as providing real-time information about the objects they are viewing.

PLASTICS MINUS PETROLEUM

Plastic is typically made from chemicals extracted from fossil fuels in a process that harms the environment. However, it is possible to make plastics from renewable biomass such as corn, sugarcane, vegetable oil, wood chips, and food waste. These are known as bioplastics. There are many types of bioplastics, some of which are biodegradable or even compostable—these are ideal for composting bags, packaging film, and other disposable uses. However, less than 1 percent of the plastic produced annually is from biomass. Researchers hope to make such materials cheaper and higher-performing so that they are able to compete with conventional plastics.

TURNING CORN INTO PLASTIC

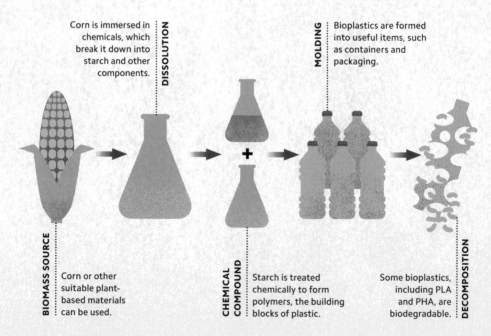

DISSOLUTION
Corn is immersed in chemicals, which break it down into starch and other components.

MOLDING
Bioplastics are formed into useful items, such as containers and packaging.

BIOMASS SOURCE
Corn or other suitable plant-based materials can be used.

CHEMICAL COMPOUND
Starch is treated chemically to form polymers, the building blocks of plastic.

DECOMPOSITION
Some bioplastics, including PLA and PHA, are biodegradable.

PLASTIC-EATING ENZYMES

With plastic pollution threatening environmental and public health, scientists are trying to find ways to break plastic down faster. It takes around 450 years for a bottle made of polyethylene terephthalate (PET) plastic to decompose. One possibility involves engineering bacteria to "eat" plastic waste. Some bacteria are able to break down plastics into harmless substances—for example, *Ideonella sakaiensis* has developed the ability to break down PET using the enzymes PETase and MHETase. It is hoped that, with genetic engineering (see p.34), such properties could be exploited to break down plastics at a useful rate.

ENGINEERING PLASTIC-EATING BACTERIA

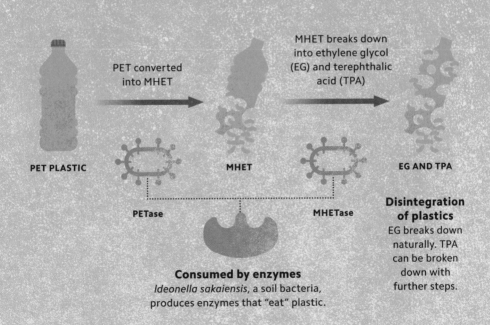

PET converted into MHET

MHET breaks down into ethylene glycol (EG) and terephthalic acid (TPA)

PET PLASTIC

MHET

EG AND TPA

PETase

MHETase

Disintegration of plastics
EG breaks down naturally. TPA can be broken down with further steps.

Consumed by enzymes
Ideonella sakaiensis, a soil bacteria, produces enzymes that "eat" plastic.

PRINTING EVERYTHING

Additive manufacturing, or 3D printing, describes a number of different processes by which 3D objects are constructed from digital models using a computer-controlled machine called a 3D printer. This reduces manufacturing costs, allows for greater customization, and can produce complex shapes with high precision. Researchers are working on printing a wide range of objects, including food (see p.55), bespoke medical implants, and even living tissue (see p.39).

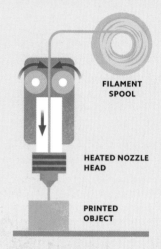

FILAMENT SPOOL

HEATED NOZZLE HEAD

PRINTED OBJECT

Extrusion
Filament (usually plastic) is melted in a nozzle head and squeezed out onto a surface.

2D object
First, a flat object is printed. Within the structure is a careful arrangement of materials that have varying properties. It is easy to store and transport.

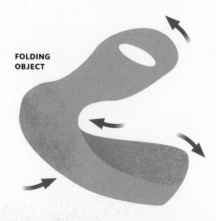

FOLDING OBJECT

Add heat and moisture
Next, the flat object is immersed in hot water. Its various materials respond in different ways to the heat and moisture, making the object fold.

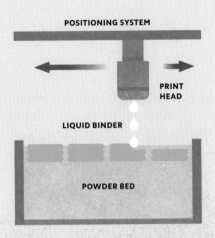

Binder jetting

The print head deposits liquid binding agent on layer after layer of powdered material to build up the object.

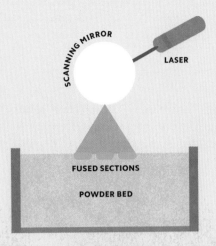

Laser sintering

A powerful laser beam is directed via a mirror onto the powder, which fuses into a solid mass.

Chair is formed

The materials are "programmed" to stop folding when they have formed a chair. Objects of all shapes and sizes can be produced in this way.

ULTIMATE ORIGAMI

In 4D printing, or active origami, additive manufacturing produces "live" 3D objects that change shape in response to external stimuli, such as heat or light. This is usually achieved by arranging materials within the object, each of which responds differently to the same stimulus, such as causing certain parts to swell in the presence of water. Uses include adaptive pipes that change diameter and self-assembling objects.

BIOTECH

NOLOGY

Groundbreaking advances in biotechnology are bringing about transformations in many fields of human activity, not least that of health care. Bioprinting is making the manufacture of human organs a real possibility. Nanoscale medicines allow for improved precision and the limiting of unwanted side effects, and laboratory functions have been shrunk to the size of a postage stamp, enabling bedside testing and instant results. Scientists can now use cellular engineering to heal diseased or damaged body parts, while genetic engineering has implications for the treatment of acute disorders as well as the revival of extinct species.

BREAKING GENETIC CODES

Genetics looks at single genes (see right). Genomics is the study of the entirety of an organism's genes: its genome. Scientists have fully sequenced the human genome (determined the order of its billions of nucleotides—the building blocks of RNA and DNA), along with those of many other organisms. This has led to biological and medical breakthroughs. For example, parts of the genome linked to desirable traits can be used to make useful varieties of another organism, such as drought-resistant crops (see p.50).

The detail

Genetics reveals how individual genes (sequences of DNA on a single chromosome) can pass on traits and conditions (see below) to new generations.

SINGLE GENE ON A CHROMOSOME

1	2	3	4	5	6
7	8	9	10	11	12
13	14	15	16	17	18
19	20	21	22	X	Y

GENE INTERACTIONS ON A COMPLETE SET OF CHROMOSOMES

The big picture

Genomics uses genome sequencing technology to study how genes relate to one another and the environment. This helps us understand how such factors influence an organism's development or contribute to disease.

+

LIFESTYLE AND ENVIRONMENTAL FACTORS

=

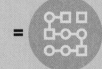

PATTERNS FOR COMPLEX DISEASE, SUCH AS HEART DISEASE OR DIABETES

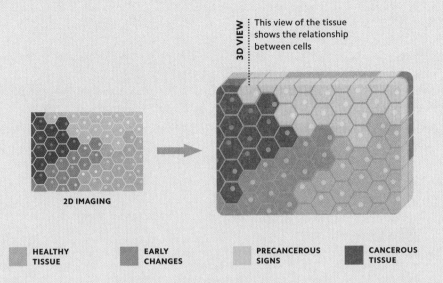

This view of the tissue shows the relationship between cells

2D IMAGING

| HEALTHY TISSUE | EARLY CHANGES | PRECANCEROUS SIGNS | CANCEROUS TISSUE |

Mapping tissue
By creating extremely detailed 3D models of tissues, spatial biologists can not only identify diseased tissue but also detect early changes in surrounding cells.

BIOLOGICAL LANDSCAPES

Biological systems are intricate 3D networks. Using spatial biology, it is now possible to create corresponding 3D maps of them. Spatial biology combines advanced genome sequencing with imaging techniques such as immunofluorescence (using fluorescent dyes to visualize targets) to model how millions of different cell types are organized in tissue. This allows the study of cells, proteins, and other factors across multiple dimensions within their complex biological landscapes. The level of detail provides unprecedented insights—for example, revealing the activity of cells within tumors (see above).

TAILORED HEALTH CARE

A person's risk of developing a particular disease, or of not responding to a treatment, can be predicted in part by the DNA sequence of their genome. Studying the links between genomics (including research into genetic biomarkers) and disease helps doctors guide patients toward tailored prevention and treatment plans. The number of people who have had their genome sequenced has grown from one at the end of the 20th century to tens of millions in the 2020s, with personalized medicine widely predicted to revolutionize health care in the years to come.

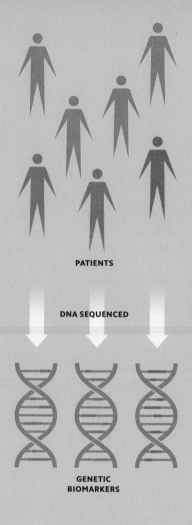

PATIENTS

DNA SEQUENCED

GENETIC BIOMARKERS

OPTIMAL TREATMENT IDENTIFIED

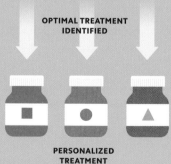

PERSONALIZED TREATMENT

Pinpointing differences
Identifying a particular genetic mutation (see p.36), coupled with information about a patient's lifestyle, tells doctors which treatment has the best chance of success.

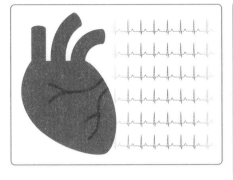

ORIGINAL HEART

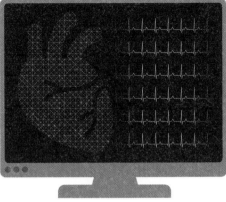

DIGITAL TWIN

Modeling the heart
Simulation software predicts the outputs from diagnostic images produced by computerized tomography (CT; see p.39) and magnetic resonance imaging (MRI) scanners, as well as molecular analyses and symptom scores. From these, it creates the digital twin of a patient's heart.

VIRTUALLY THE SAME

Digital representations, or twins, of physical objects are used to simulate real behavior. A digital twin of an organ, such as a heart, is created using continuously updated information from a living individual's heart. The twin shows doctors and scientists how the heart is malfunctioning and what treatment would be most likely to fix the problem. In the case of cardiovascular disease—such as arrhythmia—digital twins will improve the way patients with the same condition are categorized, using multiple clinical, imaging, molecular, and other variables to guide diagnosis and treatment.

AI HEALTH CHECK

Artificial intelligence (AI) has the potential to increase the accuracy and speed of medical diagnosis, freeing up time to focus on patient care. AI can process large amounts of data, including electrocardiograms (ECGs), pulse measurements, medical histories, and demographic information. This is used to build a more complete patient picture, helping medics identify risk, prevent disease, diagnose conditions earlier, and reduce misdiagnosis. AI increases treatment options and improves outcomes. Precision and the ability to identify biomarkers early on can facilitate more tailored cancer treatment, for instance.

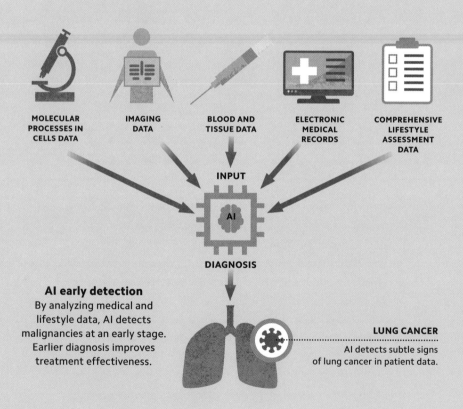

MOLECULAR PROCESSES IN CELLS DATA

IMAGING DATA

BLOOD AND TISSUE DATA

ELECTRONIC MEDICAL RECORDS

COMPREHENSIVE LIFESTYLE ASSESSMENT DATA

INPUT

AI

DIAGNOSIS

AI early detection
By analyzing medical and lifestyle data, AI detects malignancies at an early stage. Earlier diagnosis improves treatment effectiveness.

LUNG CANCER
AI detects subtle signs of lung cancer in patient data.

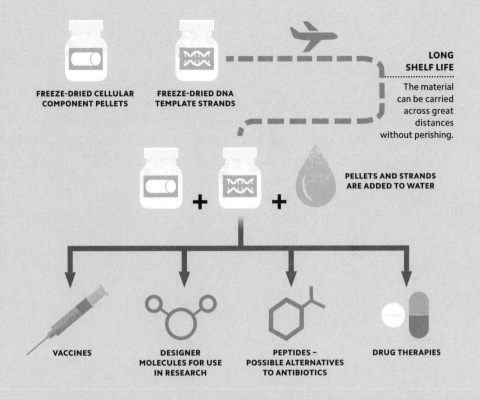

FREEZE-DRIED CELLULAR
COMPONENT PELLETS

FREEZE-DRIED DNA
TEMPLATE STRANDS

**LONG
SHELF LIFE**
The material
can be carried
across great
distances
without perishing.

**PELLETS AND STRANDS
ARE ADDED TO WATER**

VACCINES

DESIGNER
MOLECULES FOR USE
IN RESEARCH

PEPTIDES –
POSSIBLE ALTERNATIVES
TO ANTIBIOTICS

DRUG THERAPIES

READY-TO-MIX MEDICINE

The demand for biopharmaceuticals—drugs or vaccines
made from living cells or organisms—is growing. Medicines
containing DNA need to be stored at –112°F (–80°C) but are
often required far away from a lab's cold-storage unit. Freeze-
drying enables them to be safely transported and stored;
cellular components and DNA templates can be freeze-dried
separately into tiny pellets that remain stable for extended
periods. These can then be taken to low-resource settings
or war zones, for example, where they are hydrated together
to produce on-demand drugs, vaccines, and other therapies.

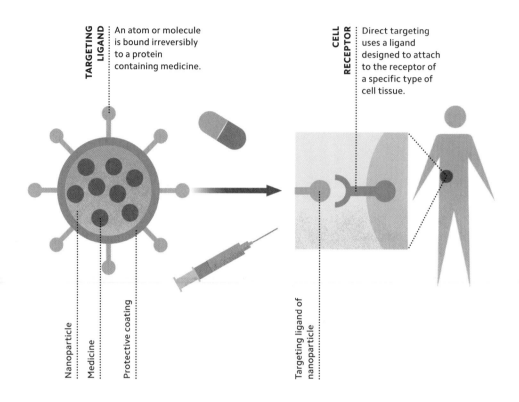

TARGETING LIGAND — An atom or molecule is bound irreversibly to a protein containing medicine.

CELL RECEPTOR — Direct targeting uses a ligand designed to attach to the receptor of a specific type of cell tissue.

Nanoparticle

Medicine

Protective coating

Targeting ligand of nanoparticle

MINIATURE MEDICINE

Nanomedicine involves the use of nanomaterials (nanoparticles) 1–100 nm in diameter. Nanoparticles alone do not respond to their surroundings or target specific sites, unlike micro- and nanobots (see p.116). Instead, nanomedicine uses ligands (atoms or molecules) that target and attach themselves to specific cells. Nanoparticles' size and relatively large surface area is associated with high therapeutic efficacy, meaning less medicine is needed and fewer side effects are expected. Nanoparticles are also used for diagnosis, monitoring, and disease prevention.

INSTANT DIAGNOSIS

A lab-on-a-chip is a tiny laboratory on a platform no bigger than a microscope slide—typically 3 sq in (19.5 sq cm). Microchannels in the platform guide volumes of sample fluid between a series of operations, making multiple analyses simultaneously on single cells and single-drop samples. Doctors can carry out these tests at a patient's bedside. The system requires integrated pumps and, like any full-size lab, valves, reagents, and electronics.

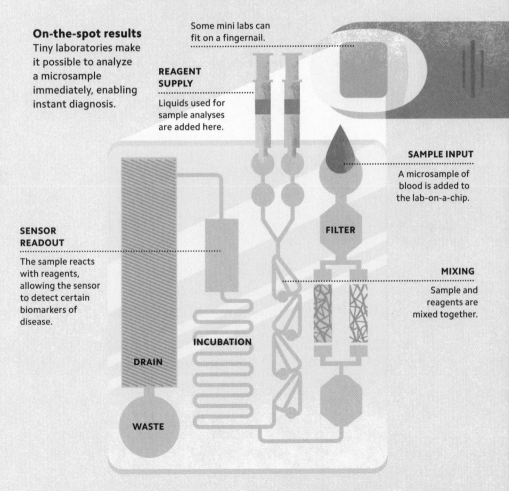

On-the-spot results
Tiny laboratories make it possible to analyze a microsample immediately, enabling instant diagnosis.

Some mini labs can fit on a fingernail.

REAGENT SUPPLY
Liquids used for sample analyses are added here.

SAMPLE INPUT
A microsample of blood is added to the lab-on-a-chip.

SENSOR READOUT
The sample reacts with reagents, allowing the sensor to detect certain biomarkers of disease.

FILTER

MIXING
Sample and reagents are mixed together.

INCUBATION

DRAIN

WASTE

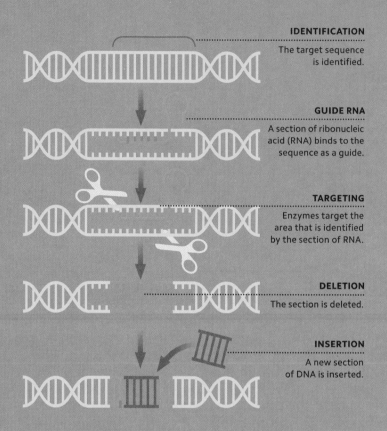

IDENTIFICATION

The target sequence
is identified.

GUIDE RNA

A section of ribonucleic
acid (RNA) binds to the
sequence as a guide.

TARGETING

Enzymes target the
area that is identified
by the section of RNA.

DELETION

The section is deleted.

INSERTION

A new section
of DNA is inserted.

CUT-AND-PASTE DNA

Scientists can manipulate genetic sequences by inserting or removing deoxyribonucleic acid (DNA) segments. A faulty gene can be removed or edited, or a gene from one species can be inserted into another species. Such genetic modification (GM) technology is used in crop farming to improve resistance to herbicides and pesticides, or to boost the production of particular nutrients. Genetic engineering is also used to develop microorganisms that produce human insulin to treat diabetes, or blood-clotting proteins to treat hemophilia. The gene-editing tool CRISPR-Cas9, which recognizes and cuts specific DNA sequences, has improved the efficiency, cost, and accuracy of gene therapy (see p.36).

ENHANCING NATURE

Synthetic biology involves redesigning organisms to perform functions that they don't carry out in nature, and which could benefit various industries. Unlike genome editing, where small changes to DNA sequences are made, here an organism's entire genetic code is altered, by inserting long stretches of synthetic DNA or DNA from another species. In this way, bacteria have been engineered to produce drugs, while silkworms have been engineered to produce stronger silk.

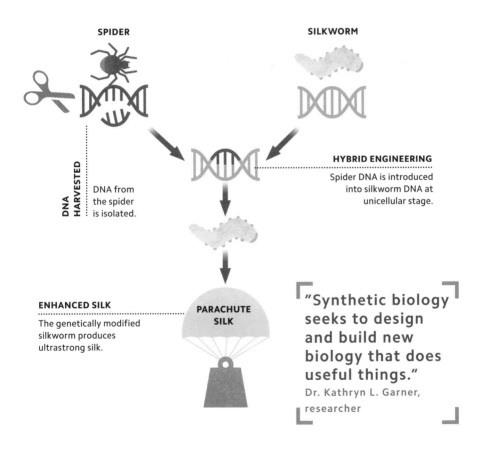

SPIDER

SILKWORM

DNA HARVESTED
DNA from the spider is isolated.

HYBRID ENGINEERING
Spider DNA is introduced into silkworm DNA at unicellular stage.

ENHANCED SILK
The genetically modified silkworm produces ultrastrong silk.

PARACHUTE SILK

"Synthetic biology seeks to design and build new biology that does useful things."
Dr. Kathryn L. Garner, researcher

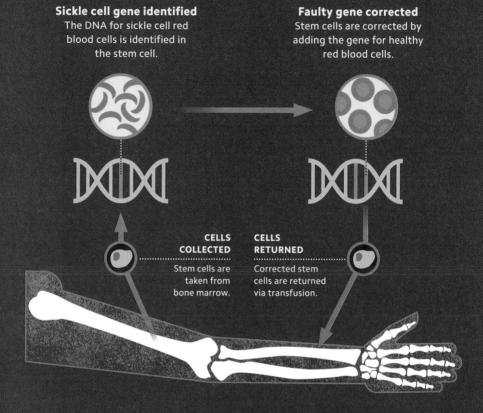

Sickle cell gene identified
The DNA for sickle cell red blood cells is identified in the stem cell.

Faulty gene corrected
Stem cells are corrected by adding the gene for healthy red blood cells.

CELLS COLLECTED
Stem cells are taken from bone marrow.

CELLS RETURNED
Corrected stem cells are returned via transfusion.

CHOPPING AND CHANGING

The prospect of treating genetic disorders by transferring genetic material into cells has become a reality. It is now possible to replace or deactivate disease-causing genes, or add modified genes to patients' cells using a variety of carriers or vectors. Bone marrow is a rich source of stem cells that can be collected from a patient and given a healthy or corrected gene. For example, scientists can remove stem cells from the bone marrow of a patient suffering from sickle cell disease, replace the disease-causing hemoglobin gene, and transfuse the amended cells back into the patient's bloodstream.

CELLULAR ENGINEERING

Cell therapy is used to halt or reverse disease and restore damaged organs. The cell types involved include stem cells, which can differentiate into almost any type of cell in the body, and cells engineered outside the body. Medics can remove a cancer patient's T-cells (white blood cells that are part of the immune system) and modify them to become chimeric antigen receptor (CAR) T-cells. When CAR T-cells are reintroduced into the patient's bloodstream, they identify specific proteins in cancer cells and destroy them.

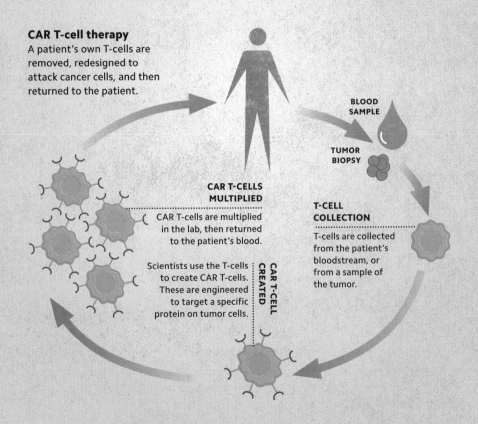

CAR T-cell therapy
A patient's own T-cells are removed, redesigned to attack cancer cells, and then returned to the patient.

BLOOD SAMPLE

TUMOR BIOPSY

CAR T-CELLS MULTIPLIED
CAR T-cells are multiplied in the lab, then returned to the patient's blood.

T-CELL COLLECTION
T-cells are collected from the patient's bloodstream, or from a sample of the tumor.

Scientists use the T-cells to create CAR T-cells. These are engineered to target a specific protein on tumor cells.

CAR T-CELL CREATED

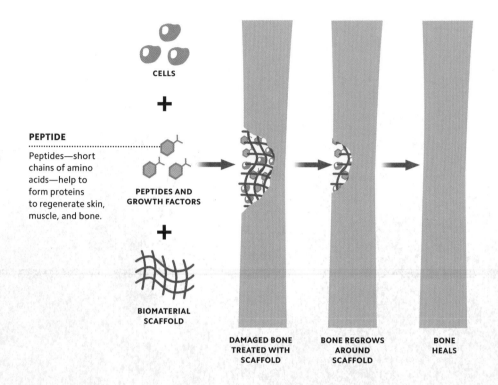

CELLS

+

PEPTIDE

Peptides—short
chains of amino
acids—help to
form proteins
to regenerate skin,
muscle, and bone.

**PEPTIDES AND
GROWTH FACTORS**

+

**BIOMATERIAL
SCAFFOLD**

**DAMAGED BONE
TREATED WITH
SCAFFOLD**

**BONE REGROWS
AROUND
SCAFFOLD**

**BONE
HEALS**

A STRUCTURED RECOVERY

By means of tissue engineering, a patient's cells can be helped
to grow around a damaged or diseased area using a scaffold made
of absorbable or biodegradable material. Cells, including stem cells,
and growth factors are added to the scaffold, where they grow and
develop into new tissue. Engineering tissue avoids the need to
transplant it from a donor. Suitable tissues include those of bone,
skin, cartilage, and heart. Cartilage has been regenerated in a damaged
knee using this technology. Scientists have also created mature bone
grafts using bone cells (osteoblasts) grown in the laboratory together
with growth factors and a biomaterial scaffold.

BESPOKE BODY PARTS

Tissues can be printed with a 3D printer using cells and biomaterials rather than ink or plastics. The printer follows instructions taken from scans of existing tissue. These scans are produced by CT (computerized tomography, which collates a series of X-ray images) and MRI (magnetic resonance imaging) machines. The resulting tissue is printed from the bottom up using a mixture of cells, matrix, and nutrients called "bioink." These 3D-printed tissues offer an improved alternative to cells grown in petri dishes. The ultimate goal is to grow entire organs for transplant.

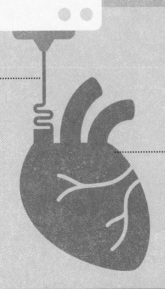

BIOINK

The ink in the bioprinting process is a mixture of cells (in this case, cardiac muscle cells), matrix, and nutrients.

3D STRUCTURE

Bioinks are 3D bioprinted into tissue constructs (engineered tissue) for drug screening, disease modeling, and in vitro transplantation.

ARTIFICIAL WOMBS

Recreating the environment in which an unborn baby develops could improve outcomes for very premature births. Researchers have transferred lamb fetuses to sterile bags that act as artificial wombs. The fetal heart needs to be sufficiently developed to pump blood. Fluid that simulates amniotic fluid (the liquid around the fetus) is pumped in and out of the bag. A replacement placenta called an "oxygenator" is connected to the fetus via the umbilical cord. The fetal heart pumps blood and waste to the placenta, and oxygenated blood and nutrients are returned.

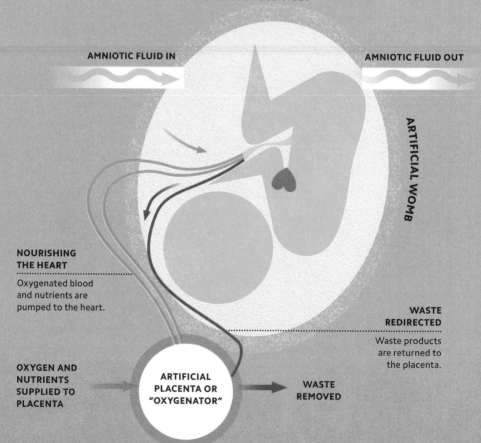

AMNIOTIC FLUID IN

AMNIOTIC FLUID OUT

ARTIFICIAL WOMB

NOURISHING THE HEART
Oxygenated blood and nutrients are pumped to the heart.

WASTE REDIRECTED
Waste products are returned to the placenta.

OXYGEN AND NUTRIENTS SUPPLIED TO PLACENTA

ARTIFICIAL PLACENTA OR "OXYGENATOR"

WASTE REMOVED

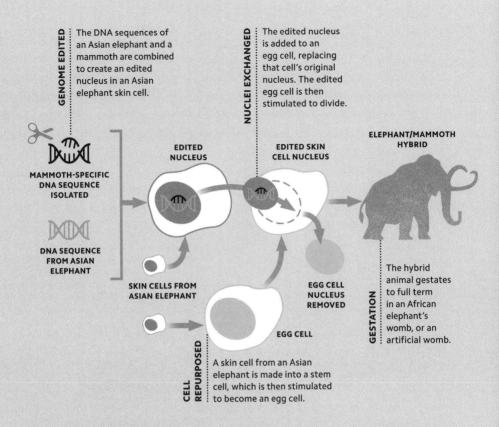

GENOME EDITED
The DNA sequences of an Asian elephant and a mammoth are combined to create an edited nucleus in an Asian elephant skin cell.

NUCLEI EXCHANGED
The edited nucleus is added to an egg cell, replacing that cell's original nucleus. The edited egg cell is then stimulated to divide.

MAMMOTH-SPECIFIC DNA SEQUENCE ISOLATED

DNA SEQUENCE FROM ASIAN ELEPHANT

EDITED NUCLEUS

EDITED SKIN CELL NUCLEUS

ELEPHANT/MAMMOTH HYBRID

SKIN CELLS FROM ASIAN ELEPHANT

EGG CELL NUCLEUS REMOVED

EGG CELL

GESTATION
The hybrid animal gestates to full term in an African elephant's womb, or an artificial womb.

CELL REPURPOSED
A skin cell from an Asian elephant is made into a stem cell, which is then stimulated to become an egg cell.

REVIVING LOST SPECIES

Biotechnology offers the possibility of creating animals that resemble extinct species. De-extinction uses methods like those used to clone Dolly the sheep in 1996: an adult cell with edited DNA is fused with an unfertilized egg without its DNA. Scientists theorize that the gene-editing method CRISPR-Cas9 (see p.34) could help reintroduce lost species, using an extinct animal's genome as a blueprint. However, the tissue from such animals may well be old and damaged, making its genome fragmented. Current discussions around reviving a mammoth could involve its gestation in an artificial womb (see opposite).

CELL RENEWAL

When young cells become diseased or harmed, they are naturally cleared from the body by a process called apoptosis (programmed cell death). However, senescent (old, decaying) cells are not cleared. They amass with age and at sites of chronic disorders, causing damage. Researchers have trialed so-called senolytic drugs, which clear senescent cells and induce apoptosis, to treat arthritis and several other inflammatory disorders.

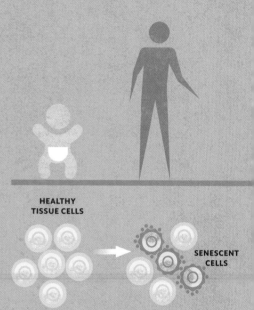

HEALTHY TISSUE CELLS

SENESCENT CELLS

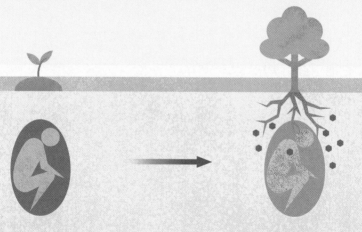

Burial and planting
The body is placed in a biodegradable pod and buried deep in the ground. A tree is planted in the soil above the pod.

Decomposition
Over a period of months, the pod breaks down, enabling the organic matter it contains to transform into mineral nutrients that support plant growth.

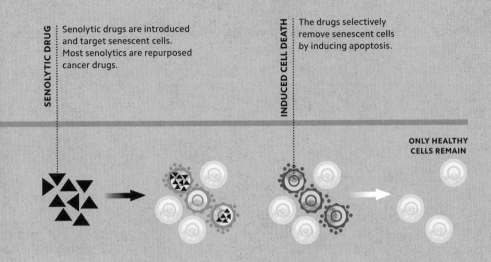

Senolytic drugs are introduced and target senescent cells. Most senolytics are repurposed cancer drugs.

The drugs selectively remove senescent cells by inducing apoptosis.

ONLY HEALTHY
CELLS REMAIN

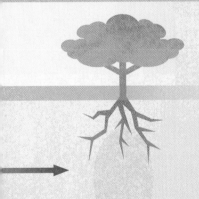

ECO-FRIENDLY FAREWELLS

Biodegradable burial pods offer one option for more environmentally friendly interment. Made from starch-based bioplastics, they are planted beneath a tree, sapling, or seed. When the pods break down, they release organic matter that transforms into minerals to help nourish the tree. Other sustainable burial practices include aquamation (liquefying in alkaline solution), recomposing (human composting), and fungal funerals.

Decomposition
Over a period of months, the pod breaks down, enabling the organic matter it contains to transform into mineral nutrients that support plant growth.

FOOD AND FARMIN

G

Today's farmers face two conflicting challenges: feeding a growing global population and cutting greenhouse gas emissions. New technologies can assist with both, especially by helping farmers make the most efficient use of their resources. This can include growing crops in indoor "vertical farms," using real-time data collected across farms to inform decision-making, and deploying multipurpose agricultural machines. Genetic engineering might have a major role to play in adapting cereal crops to grow in adverse conditions. Meanwhile, efforts are underway to develop new, more sustainable types of food, including microalgae, lab-grown meat, and 3D-printed meals.

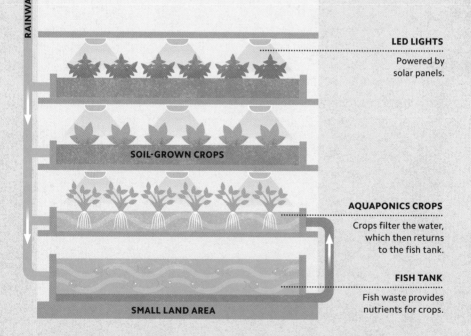

SOLAR PANELS

RAINWATER COLLECTION TANK

CROPS IN THE SKY

Responsible land use is a major element in mitigating climate change, protecting nature, and supporting communities. Vertical farming aims to minimize the area needed for agriculture by stacking crops in layers indoors, such as in greenhouses, warehouses, or former mines. Crops are given optimal conditions for growth, with light and temperature carefully controlled. However, vertical farming is strongly affected by energy prices and is not yet cost-effective compared with traditional farming.

RAINWATER

LED LIGHTS
Powered by
solar panels.

SOIL-GROWN CROPS

AQUAPONICS CROPS
Crops filter the water,
which then returns
to the fish tank.

FISH TANK
Fish waste provides
nutrients for crops.

SMALL LAND AREA

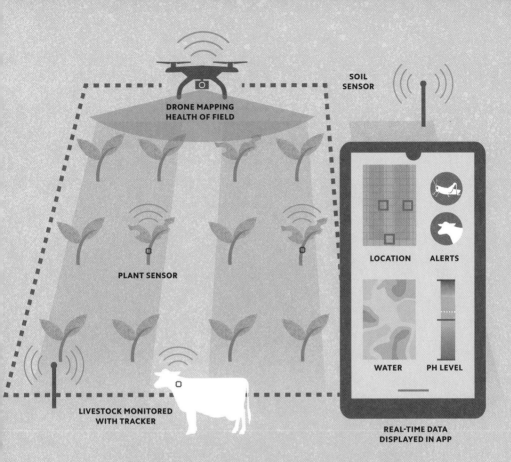

DRONE MAPPING
HEALTH OF FIELD

SOIL
SENSOR

PLANT SENSOR

LOCATION   ALERTS

WATER   PH LEVEL

LIVESTOCK MONITORED
WITH TRACKER

REAL-TIME DATA
DISPLAYED IN APP

INTERNET OF CROPS

Precision farming uses technology to monitor conditions on a farm
to improve efficiency and inform decision-making. Data from
connected devices, such as soil monitors, plant sensors, livestock
alarms, and surveillance drones, is continuously transmitted and
analyzed. On the most technologically advanced farms, this can involve
triggering automated responses. This approach aims to use resources
(such as grain, fertilizer, water, and land) as efficiently as possible.
Proponents of precision agriculture argue that a transition to this
kind of farming is necessary to sustain a growing population
without causing irreversible damage to nature.

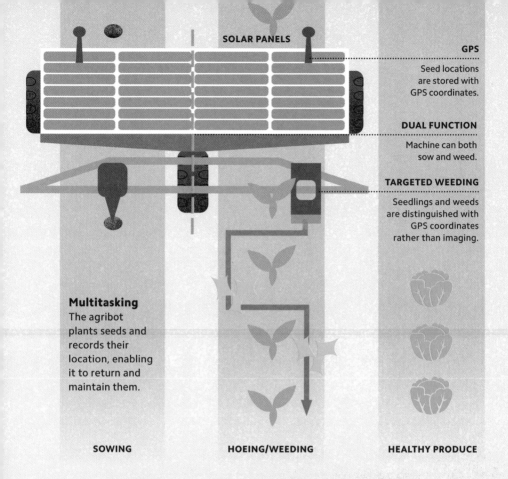

SOLAR PANELS

GPS

Seed locations are stored with GPS coordinates.

DUAL FUNCTION

Machine can both sow and weed.

TARGETED WEEDING

Seedlings and weeds are distinguished with GPS coordinates rather than imaging.

Multitasking
The agribot plants seeds and records their location, enabling it to return and maintain them.

SOWING

HOEING/WEEDING

HEALTHY PRODUCE

ROBOT FARMHANDS

Advances in robotics, computer vision, and related fields mean that a growing number of repetitive and laborious agricultural tasks can be assigned to autonomous machines. These devices can plow, plant seeds, pull weeds, apply water and fertilizer, and harvest crops with minimal or no human supervision. For example, a robot can store the location of the seeds it plants using GPS, so when it returns to the same spot, it can pull out weeds or precisely apply pesticides while avoiding the young crops. This process is cheaper and more environmentally friendly than spraying pesticides.

BUILT-IN FERTILIZERS

One of the key nutrients for plant growth is nitrogen. Artificial nitrogen fertilizers are critical in increasing food production to support the growing world population. However, their production process is energy intensive and usually requires fossil fuels, which are linked to environmental damage such as water pollution. There is an urgent need to find sustainable alternatives. One innovative approach is to use genetic engineering (see p. 34) to give cereal crops the ability (mostly limited to microorganisms at present) to "fix" nitrogen—that is, to convert atmospheric nitrogen into useful nitrogen compounds.

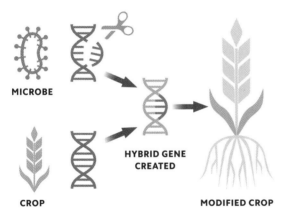

MICROBE

CROP

HYBRID GENE
CREATED

MODIFIED CROP

Nif genes
Nitrogen fixation ("nif") genes from various bacteria have been transferred to crops and "expressed"—the information in the genes is used to create molecules, such as proteins, in other organisms.

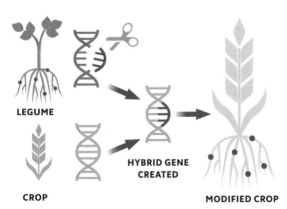

LEGUME

CROP

HYBRID GENE
CREATED

MODIFIED CROP

Root nodules
Legumes, in partnership with nitrogen-fixing bacteria, can convert atmospheric nitrogen into nutritious compounds in their root nodules. This gene can be introduced into other crops.

"STAY GREEN"

A gene in drought-resistant plants helps them stay green for longer. This prolongs the harvest season, giving a higher yield.

EARLY FLOWERING

Crops could limit damage from extreme weather by flowering early.

ACTIVE STOMATA

Transgenic maize with more active stomata (pores that can close to control water loss) is more drought tolerant.

WAX COVERING

Plants that have more protective wax are more likely to survive droughts, and the damaging effects of UV light and severe cold.

FLOWER POWER

Transferring a floral trait to corn affects the way it uses carbon, altering its growth pattern and increasing yield.

Transgenic crops

By transferring genes associated with drought tolerance into cereal crops such as corn (shown), the plant's resistance to drought would be boosted.

ROOT ARCHITECTURE

Gene manipulation may improve root growth in droughts so that the plant can seek out new water sources.

WHAT DROUGHT?

Many extreme weather events, such as drought, are becoming increasingly frequent and severe and can be devastating for crops. In an effort to improve crop yields under these difficult conditions, scientists are studying the properties of naturally drought-resistant plants with a view to engineering more resilient versions of staple foods such as wheat, rice, and corn.

Nanoparticles
These can be sprayed onto crops or added to the soil to alleviate salt stress.

Increases chlorophyll and promotes efficient photosynthesis.

Improves the activity of enzymes (proteins that speed up chemical reactions), which help seed germination.

Improves cells' ability to retain potassium, an essential nutrient.

NANOPARTICLES
ADDED TO SOIL

COTTON PLANT

Boosts molecules called antioxidants to neutralize free radicals (unstable atoms that harm cells).

Helps prevent the buildup of salt in leaves under salt stress conditions.

Assists in maintaining homeostasis (stability and self-regulation), to support plant growth.

THE SALT SOLUTION

Factors such as rising sea levels have led to a rapid increase in land affected by salinity (the presence of salt). This can cause salt stress in crops, reducing yields and threatening food supplies. Nanotechnology and genetic engineering can help crops tolerate salinity. Scientists use nanotechnology to create biocompatible nanoparticles (see p.10) that can enhance crops' nutrient uptake and water balance, among other defense measures. Nanoparticles of chitosan (a sugar) can, for instance, improve the efficiency of water use in corn, reducing salt accumulation in the plant.

ALGAL AGRICULTURE

Microalgae (single-celled, water-dwelling organisms) have, until now, been a mostly untapped natural resource. Rich in nutrients and fast growing, they can be used as ingredients in food supplements for humans and in livestock feed, fuel, fertilizer, and pharmaceuticals. Cultivation costs, particularly for harvesting, are prohibitively expensive, however. The challenge is to cultivate microalgae in a cost-effective way.

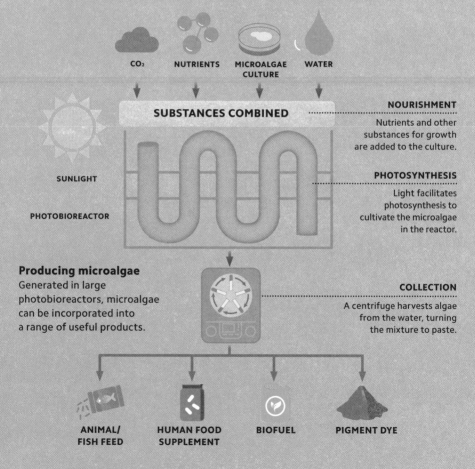

CO₂   NUTRIENTS   MICROALGAE CULTURE   WATER

SUBSTANCES COMBINED

NOURISHMENT
Nutrients and other substances for growth are added to the culture.

SUNLIGHT

PHOTOSYNTHESIS
Light facilitates photosynthesis to cultivate the microalgae in the reactor.

PHOTOBIOREACTOR

Producing microalgae
Generated in large photobioreactors, microalgae can be incorporated into a range of useful products.

COLLECTION
A centrifuge harvests algae from the water, turning the mixture to paste.

ANIMAL/ FISH FEED   HUMAN FOOD SUPPLEMENT   BIOFUEL   PIGMENT DYE

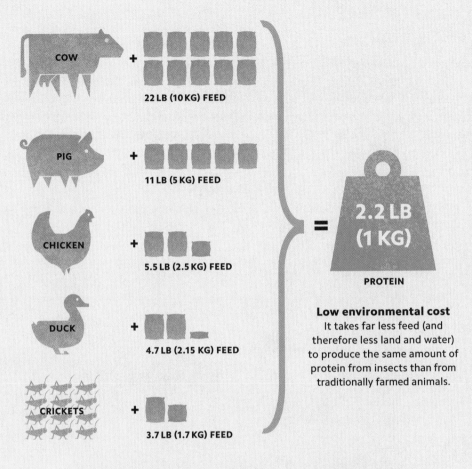

COW

\+ 22 LB (10 KG) FEED

PIG

\+ 11 LB (5 KG) FEED

CHICKEN

\+ 5.5 LB (2.5 KG) FEED

DUCK

\+ 4.7 LB (2.15 KG) FEED

CRICKETS

\+ 3.7 LB (1.7 KG) FEED

= **2.2 LB (1 KG)**

PROTEIN

Low environmental cost
It takes far less feed (and therefore less land and water) to produce the same amount of protein from insects than from traditionally farmed animals.

LOCUSTS FOR LUNCH?

Billions of people already eat insects daily. The greenhouse gas emissions and land use associated with livestock farming are now widening insects' culinary appeal in regions that do not have this tradition. Many insects, such as crickets, are highly nutritious: some contain more protein per gram than beef, pork, or chicken. They can be eaten whole or pulverized into a powder for "cricket flour," protein bars, or burger patties.

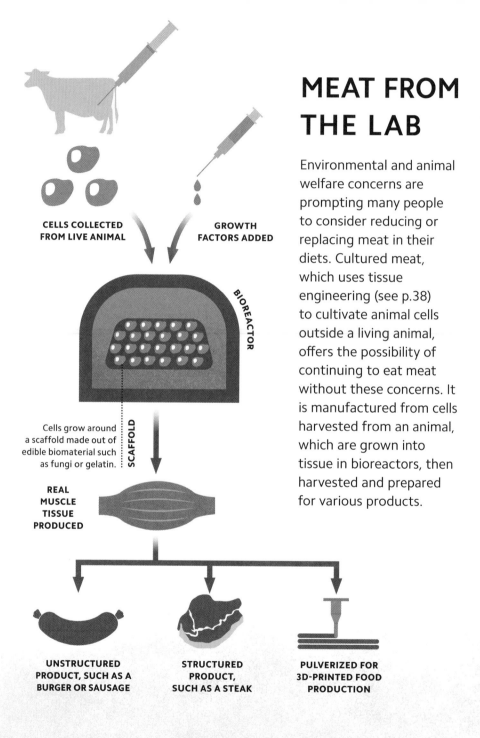

MEAT FROM THE LAB

Environmental and animal welfare concerns are prompting many people to consider reducing or replacing meat in their diets. Cultured meat, which uses tissue engineering (see p.38) to cultivate animal cells outside a living animal, offers the possibility of continuing to eat meat without these concerns. It is manufactured from cells harvested from an animal, which are grown into tissue in bioreactors, then harvested and prepared for various products.

CELLS COLLECTED FROM LIVE ANIMAL

GROWTH FACTORS ADDED

BIOREACTOR

Cells grow around a scaffold made out of edible biomaterial such as fungi or gelatin.

SCAFFOLD

REAL MUSCLE TISSUE PRODUCED

UNSTRUCTURED PRODUCT, SUCH AS A BURGER OR SAUSAGE

STRUCTURED PRODUCT, SUCH AS A STEAK

PULVERIZED FOR 3D-PRINTED FOOD PRODUCTION

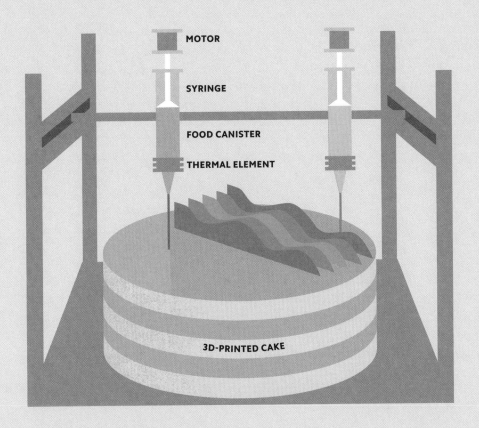

MOTOR

SYRINGE

FOOD CANISTER

THERMAL ELEMENT

3D-PRINTED CAKE

PRINTED CUISINE

All sorts of materials can be used in additive manufacturing (see p. 22), including some foodstuffs. Conventional 3D printers can be adapted to extrude molten chocolate, icing, dough, mashed potato, cultured meat, and other pastelike foodstuffs, following a digital design with single or multiple ingredients. Often, the food is heated in the machine to soften it and later cooled when it is dispensed onto the print bed. This process allows for precise control over the structure and nutritional content of the final product, with potential applications for health care and crewed space exploration.

TRANSP

ORT

A fifth of global CO₂ emissions are caused by transportation, and there is a huge effort under way to make the sector greener. This means having fewer fossil fuel-burning vehicles and more of those powered by electricity or more sustainable fuels. The push to sustainability also involves using technology to connect people and transportation, making the most efficient use of vehicles, fuel, and infrastructure, and enabling a shift away from unnecessary vehicle ownership. Vehicles of all kinds are becoming smarter—collecting, analyzing, and sharing data to make trips faster and safer. Self-driving cars are appearing on our roads, while aerial and underwater drones are taking on a wider range of jobs.

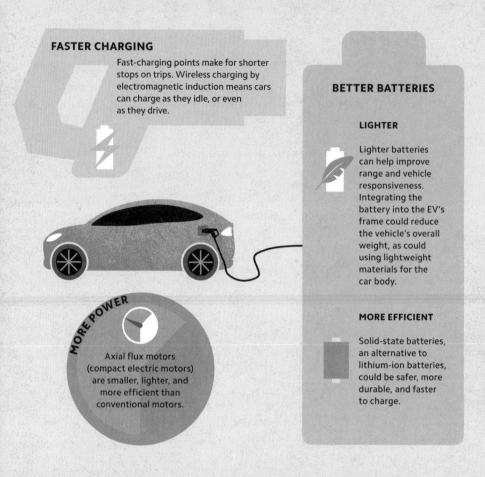

FASTER CHARGING

Fast-charging points make for shorter stops on trips. Wireless charging by electromagnetic induction means cars can charge as they idle, or even as they drive.

BETTER BATTERIES

LIGHTER

Lighter batteries can help improve range and vehicle responsiveness. Integrating the battery into the EV's frame could reduce the vehicle's overall weight, as could using lightweight materials for the car body.

MORE EFFICIENT

Solid-state batteries, an alternative to lithium-ion batteries, could be safer, more durable, and faster to charge.

MORE POWER

Axial flux motors (compact electric motors) are smaller, lighter, and more efficient than conventional motors.

EFFICIENT VEHICLES

Cars that run on polluting fossil fuels are being phased out in many markets, putting pressure on car manufacturers to provide practical, affordable, and attractive electric vehicles (EVs). There are many challenges in making EVs that can compete with their conventional counterparts, the main one being that fossil fuels are far more energy dense—that is, they provide more energy relative to their volume—than even the best lithium-ion batteries. Engineers are working to optimize EVs to ensure that they can perform as well as fossil fuel-powered cars.

FUELED BY SUNLIGHT

Solar vehicles—EVs powered by sunlight—include cars, buses, trains, aircraft, spacecraft, and boats. A typical solar-powered ground vehicle has roof-mounted solar panels that convert the sun's energy into electric energy for propulsion and auxiliary features, such as communication. Harvesting sufficient energy to run such EVs is a huge challenge, with most solar vehicles still in the research and development phase, although several solar boats are now commercially available.

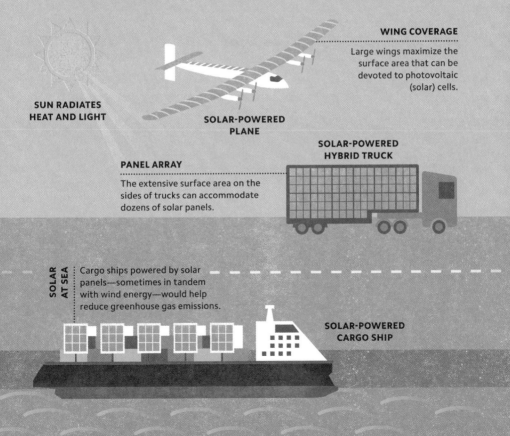

WING COVERAGE

Large wings maximize the surface area that can be devoted to photovoltaic (solar) cells.

SUN RADIATES HEAT AND LIGHT

SOLAR-POWERED PLANE

SOLAR-POWERED HYBRID TRUCK

PANEL ARRAY

The extensive surface area on the sides of trucks can accommodate dozens of solar panels.

SOLAR AT SEA

Cargo ships powered by solar panels—sometimes in tandem with wind energy—would help reduce greenhouse gas emissions.

SOLAR-POWERED CARGO SHIP

GUILT-FREE FLYING?

Developing sustainable alternatives to aviation fuel is a major hurdle for the aviation sector. Although it is relatively cheap and energy dense, kerosene-based aviation fuel is obtained from petroleum. Sustainable aviation fuels (SAFs) blend kerosene with other fuels that are chemically similar but made from sustainable sources and processes. The hope is to reduce the kerosene in SAFs, and ultimately to power large aircraft by electricity or by renewable, zero-carbon fuels such as "synthetic kerosene" derived from green hydrogen.

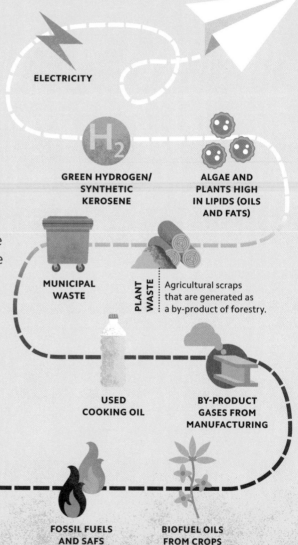

ELECTRICITY

GREEN HYDROGEN/ SYNTHETIC KEROSENE

ALGAE AND PLANTS HIGH IN LIPIDS (OILS AND FATS)

MUNICIPAL WASTE

PLANT WASTE
Agricultural scraps that are generated as a by-product of forestry.

USED COOKING OIL

BY-PRODUCT GASES FROM MANUFACTURING

FOSSIL FUELS
These energy sources are nonrenewable and harmful to the environment.

FOSSIL FUELS AND SAFS

BIOFUEL OILS FROM CROPS

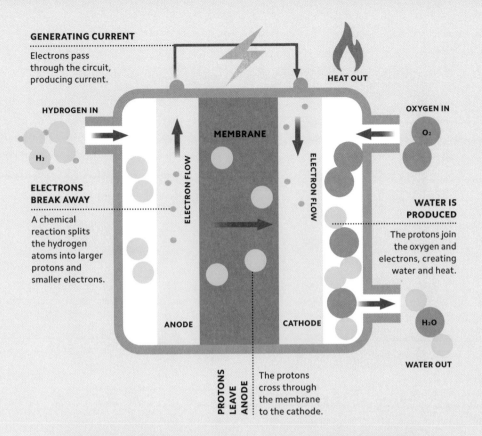

GENERATING CURRENT

Electrons pass through the circuit, producing current.

HEAT OUT

HYDROGEN IN

H₂

MEMBRANE

OXYGEN IN

O₂

ELECTRON FLOW

ELECTRON FLOW

ELECTRONS BREAK AWAY

A chemical reaction splits the hydrogen atoms into larger protons and smaller electrons.

WATER IS PRODUCED

The protons join the oxygen and electrons, creating water and heat.

ANODE

CATHODE

H₂O

WATER OUT

PROTONS LEAVE ANODE

The protons cross through the membrane to the cathode.

POWERED BY HYDROGEN

Fuel cells use chemical reactions to convert a fuel's chemical energy to electrical energy. Hydrogen fuel cells could play a key part in a net-zero-carbon-emissions future—for example, as an alternative to combustion engines. In this type of fuel cell, hydrogen atoms enter the anode (negative electrode) and are stripped of their electrons, which flow through a circuit—producing an electrical current—to the cathode (positive electrode). The protons left behind pass through a membrane to the cathode, joining oxygen to produce water and heat. Hydrogen fuel cells are efficient and versatile, with applications including transportation and power grids. However, high cost and lack of hydrogen infrastructure has impeded their adoption.

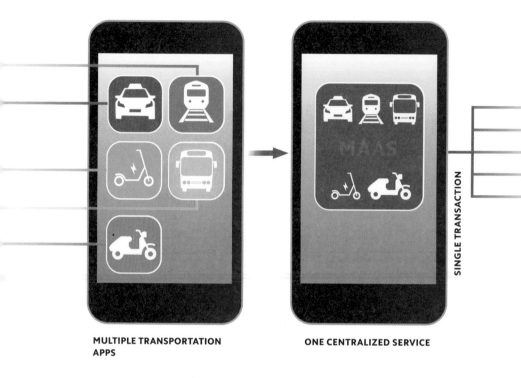

SINGLE TRANSACTION

MULTIPLE TRANSPORTATION APPS

ONE CENTRALIZED SERVICE

THE END OF CAR OWNERSHIP?

Mobility as a service (MaaS) replaces the traditional model of private vehicle ownership with a more efficient model in which people pay to access whichever transportation services they need to complete a trip. MaaS brings together public and private services, which can be planned, booked, and paid for using a single app. Users could pay for individual trips or opt to subscribe monthly to transportation services within a certain area.

ROBOTIC VEHICLES

All sorts of vehicles can be equipped with self-driving capabilities, from long-haul trucks to uncrewed minesweepers and self-landing planes. These vehicles have varying degrees of autonomy, from providing limited assistance to the driver (for instance, with steering or speed control) to operating with no human input at all. AI technology analyzes real-time data collected from sensors on the vehicle and uses it to respond to its changing environment, such as by braking when a pedestrian steps onto the road.

Self-driving truck
Many countries have a shortage of truck drivers. This has prompted growing interest in automation (or part-automation).

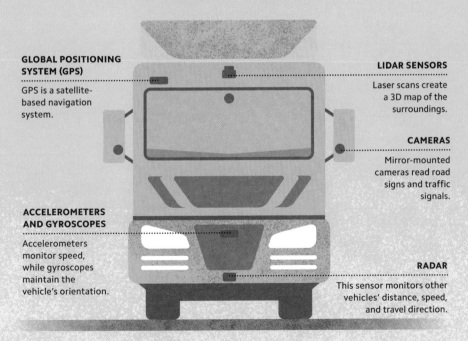

GLOBAL POSITIONING SYSTEM (GPS)

GPS is a satellite-based navigation system.

ACCELEROMETERS AND GYROSCOPES

Accelerometers monitor speed, while gyroscopes maintain the vehicle's orientation.

LIDAR SENSORS

Laser scans create a 3D map of the surroundings.

CAMERAS

Mirror-mounted cameras read road signs and traffic signals.

RADAR

This sensor monitors other vehicles' distance, speed, and travel direction.

FLOCKS OF VEHICLES

Platooning involves a group of vehicles being driven together, usually by the operator of the vehicle at the front of the platoon. This allows many vehicles to coordinate their acceleration and braking and maintain a set distance from each other, increasing the capacity of the road while reducing risk of accidents. In the future, connected cars (see below) could have the option to automatically join and leave platoons.

JOINING PLATOON
A car joins the platoon and signals its destination. The lead truck takes over the car's onboard system.

CAR SIGNALS DESTINATION

CAR PREPARES TO LEAVE PLATOON

⌐ "The technology could detect a potential collision before the driver can see the threat." ¬
Los Angeles Times

COORDINATION

Cars can continue using both lanes because they will coordinate when they reach the obstruction.

DECELERATION

This car is aware of the breakdown and decelerates.

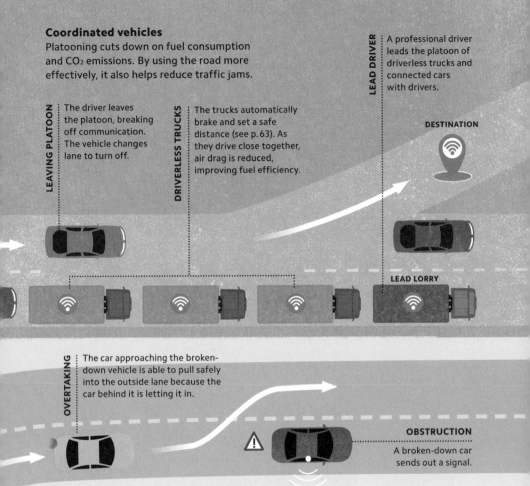

Coordinated vehicles
Platooning cuts down on fuel consumption and CO_2 emissions. By using the road more effectively, it also helps reduce traffic jams.

LEAD DRIVER
A professional driver leads the platoon of driverless trucks and connected cars with drivers.

LEAVING PLATOON
The driver leaves the platoon, breaking off communication. The vehicle changes lane to turn off.

DRIVERLESS TRUCKS
The trucks automatically brake and set a safe distance (see p. 63). As they drive close together, air drag is reduced, improving fuel efficiency.

DESTINATION

LEAD LORRY

OVERTAKING
The car approaching the broken-down vehicle is able to pull safely into the outside lane because the car behind it is letting it in.

OBSTRUCTION
A broken-down car sends out a signal.

TALKING CARS

As more cars join the Internet of Things (see p. 96), there is likely to be a growth in wireless data exchange between them. This is known as vehicle-to-vehicle (V2V) communication. Cars could transmit data about their velocity and direction, and share alerts with each other about breakdowns or dangerous weather conditions, and so improve road safety and traffic congestion.

AS FAST AS FLYING?

There are efforts to design and build rail networks that allow for hyper-fast trains, usually to link major cities. The fastest trains can travel at hundreds of miles per hour, arguably rivaling aviation for travel time and convenience. Various methods have been proposed for making them even faster. These include running trains through large vacuum tubes, thereby minimizing air drag and potentially enabling them to reach supersonic speeds with relatively little power.

CAPSULE
Each train comprises several pressurized stand-alone capsules.

COMPRESSOR FAN
The fan transfers high-pressure air from the front to the rear, creating an air cushion.

VACUUM TUBE
The sealed tube has had almost all of the air removed from it to create a near vacuum, reducing drag.

MAGNETS
Levitated by magnets, the capsule floats in air, making it frictionless.

MAGNETIC FIELD
The magnetic field propels the train forward.

DROPPING IN THE SHOPPING

Today, drones are being used all over the world to deliver goods. They are especially suited to small, urgent deliveries of medical supplies, as they do not run the risk of being held up by traffic or poor road conditions. However, they are also increasingly being employed for everyday deliveries. Releasing fleets of drones from local distribution centers to carry out the final stage in the delivery process (known as the "last mile") can reduce costs and the transportation emissions associated with distribution.

Last-mile deliveries
The "last mile," which currently makes up half of the total cost of distribution, could be performed by delivery drones.

NAVIGATE VIA GPS, SONAR, OR SIMILAR SYSTEMS

Autonomous delivery drones are equipped with satellite navigation and a suite of sensors (such as optical cameras) to enable them to arrive safely at their destination.

DELIVERY

Drone either lands with or drops the package via parachute from 200–400 ft (60–120 m) and monitors its descent.

FLEET OF DRONES RELEASED

SAFE LANDING

Customers consent to having drones drop off packages at their home, usually at the rear of their property.

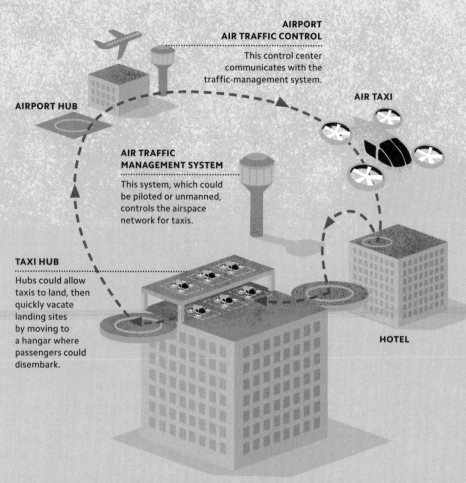

AIRPORT AIR TRAFFIC CONTROL

This control center communicates with the traffic-management system.

AIRPORT HUB

AIR TAXI

AIR TRAFFIC MANAGEMENT SYSTEM

This system, which could be piloted or unmanned, controls the airspace network for taxis.

TAXI HUB

Hubs could allow taxis to land, then quickly vacate landing sites by moving to a hangar where passengers could disembark.

HOTEL

FLYING TAXIS

There is currently a revival of interest in the concept of flying taxis (or "air taxis"), not least because they would alleviate road congestion. Dozens of companies are developing electric versions—effectively drones scaled up to carry passengers—that are quieter and less polluting than helicopters. Able to take off and land vertically, they are well suited to dense urban environments. Technical and regulatory hurdles still need to be overcome, such as the development of cheaper, lighter batteries and mechanisms to ensure safety in urban airspace.

READY TO (RE)LAUNCH

Traditionally, rockets were extremely expensive vehicles, built and used for only one launch. Reusable rockets, however, have parts (such as engines and boosters) that can be recovered, refurbished, and relaunched. This makes it cheaper to put passengers and cargo into orbit. Recyclable rockets and other reusable launch vehicles play an important role in ushering in the "New Space Age," in which space transportation is increasingly affordable and commercial, as opposed to being accessible only to a few government agencies.

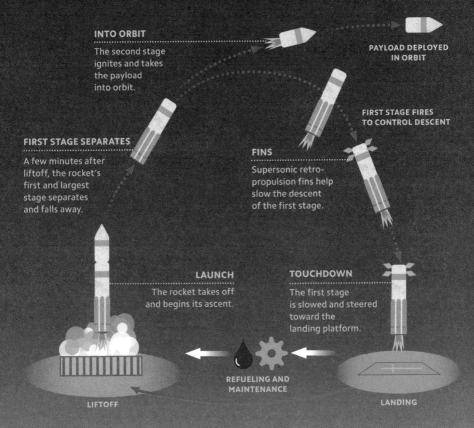

INTO ORBIT

The second stage ignites and takes the payload into orbit.

PAYLOAD DEPLOYED IN ORBIT

FIRST STAGE FIRES TO CONTROL DESCENT

FIRST STAGE SEPARATES

A few minutes after liftoff, the rocket's first and largest stage separates and falls away.

FINS

Supersonic retro-propulsion fins help slow the descent of the first stage.

LAUNCH

The rocket takes off and begins its ascent.

TOUCHDOWN

The first stage is slowed and steered toward the landing platform.

REFUELING AND MAINTENANCE

LIFTOFF

LANDING

SPACEPLANE

Spaceplanes can fly and glide like an aircraft in Earth's atmosphere and also maneuver beyond it like a spacecraft. The most famous spaceplane, NASA's Space Shuttle, entered orbit with a crew and assisted in the construction of the International Space Station. So far, all orbital spaceplanes have launched vertically on a separate rocket. Other models include a spaceplane that takes off horizontally, such as Radian One, and a small, uncrewed vehicle inside a vertical-launch rocket, such as the X37B. Both could land on an airport runway.

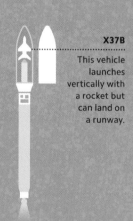

X37B

This vehicle launches vertically with a rocket but can land on a runway.

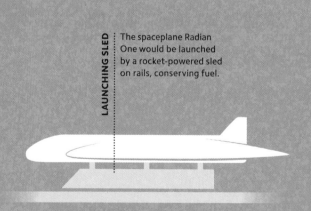

LAUNCHING SLED

The spaceplane Radian One would be launched by a rocket-powered sled on rails, conserving fuel.

UNDERWATER AUTOMATA

Autonomous submarines—robots that travel underwater without constant human supervision—have long been valuable for search and rescue, oil exploration, scientific research, and defense activities. Now, defense agencies are supporting the development of larger, tougher, more sophisticated craft that could play a leading role in the future of warfare. These submarines can carry heavier payloads (including torpedoes or missiles), dive deeper, and run for months on end, taking on tasks that previously required the presence of a crew.

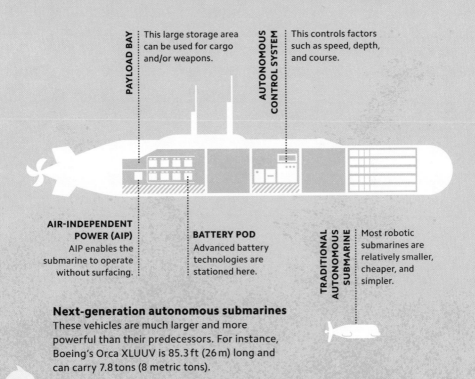

PAYLOAD BAY
This large storage area can be used for cargo and/or weapons.

AUTONOMOUS CONTROL SYSTEM
This controls factors such as speed, depth, and course.

AIR-INDEPENDENT POWER (AIP)
AIP enables the submarine to operate without surfacing.

BATTERY POD
Advanced battery technologies are stationed here.

TRADITIONAL AUTONOMOUS SUBMARINE
Most robotic submarines are relatively smaller, cheaper, and simpler.

Next-generation autonomous submarines
These vehicles are much larger and more powerful than their predecessors. For instance, Boeing's Orca XLUUV is 85.3 ft (26 m) long and can carry 7.8 tons (8 metric tons).

INFORM
TECHNO

ATION
LOGY

There is more to the future of computing than exponentially more powerful hardware. "Serverless" computing marks a shift away from centralized, inflexible IT systems, and more tasks are being automated as artificial intelligence becomes prevalent. Traditional, or "classical", silicon-based computers are unlikely to be entirely displaced but could soon feature hardware that helps meet demand for computing resources, such as optical computers that run a million times faster, or DNA storage that saves data for billions of years. Meanwhile, quantum technologies are giving rise to unbreakable encryption, ultra-precise sensors, and computers that could solve "impossible" problems.

EMULATING THE BRAIN

Artificial intelligence (AI) refers to the ability of machines to simulate intelligence. AI can be used to perform tasks that traditionally would require human intellect, such as decision-making, translation, and image recognition. Today, AI is dominated by machine learning (see opposite)—

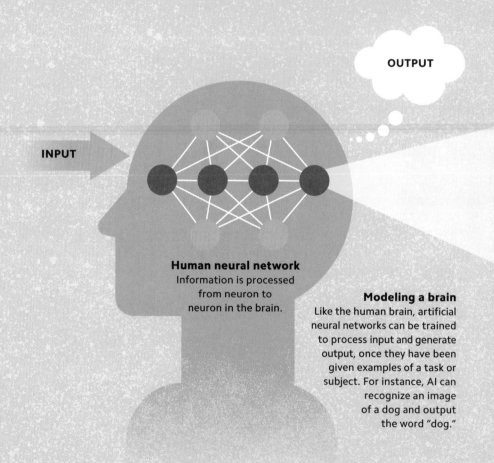

INPUT

OUTPUT

Human neural network
Information is processed
from neuron to
neuron in the brain.

Modeling a brain
Like the human brain, artificial
neural networks can be trained
to process input and generate
output, once they have been
given examples of a task or
subject. For instance, AI can
recognize an image
of a dog and output
the word "dog."

an approach that seeks to help computers "learn" how to carry out tasks without explicitly programming them to do so. The availability of vast quantities of data and computing power for training has enabled AI to take leaps forward in recent years (see pp.76–77).

ARTIFICIAL NEURON

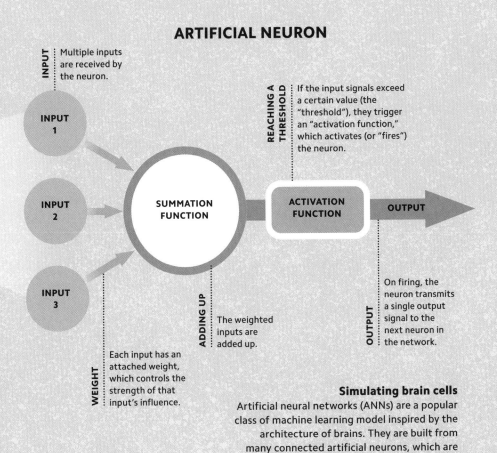

INPUT
Multiple inputs are received by the neuron.

INPUT 1

INPUT 2

INPUT 3

REACHING A THRESHOLD
If the input signals exceed a certain value (the "threshold"), they trigger an "activation function," which activates (or "fires") the neuron.

SUMMATION FUNCTION

ACTIVATION FUNCTION

OUTPUT

ADDING UP
The weighted inputs are added up.

WEIGHT
Each input has an attached weight, which controls the strength of that input's influence.

OUTPUT
On firing, the neuron transmits a single output signal to the next neuron in the network.

Simulating brain cells
Artificial neural networks (ANNs) are a popular class of machine learning model inspired by the architecture of brains. They are built from many connected artificial neurons, which are modeled on biological neurons.

AI EVERYWHERE

When they are provided with sufficient quantity and quality of training data, artificial neural networks (ANNs) and other AI models can be taught to perform certain tasks as well as—or even better than—humans. These models are now so prevalent that most people with access to computers use them every day without realizing—when using a search engine, for example. Although AI is a general-purpose technology with many mundane uses, exciting new applications are being discovered all the time. For instance, in 2021 it was revealed that an ANN called AlphaFold was able to predict the 3D structures of nearly every known protein.

The architecture of ANNs

ANNs are made up of artificial neurons organized into multiple layers. Signals begin at the input layer and travel via "hidden" layers to the output layer.

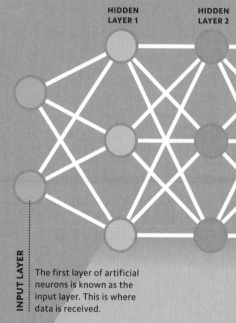

HIDDEN LAYER 1

HIDDEN LAYER 2

INPUT LAYER

The first layer of artificial neurons is known as the input layer. This is where data is received.

RESEARCH

DIAGNOSIS AND HEALTH MONITORING

PREDICTIVE MAINTENANCE

PERSONALIZED WEB FEED

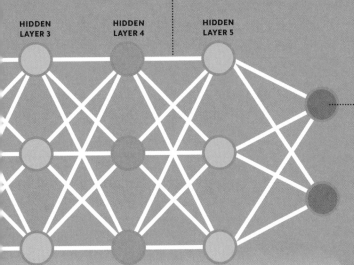

HIDDEN LAYERS

Each "hidden" layer consists of connected neurons that receive data, process it, and send it to the next layer. Multiple hidden layers enable ANNs to learn complex relationships in the data. Deep learning models use many hidden layers.

HIDDEN LAYER 3

HIDDEN LAYER 4

HIDDEN LAYER 5

OUTPUT LAYER

Transformed by its passage through layers of neurons, the processed data finally reaches the output layer, where useful results are generated.

AI USES

AI ROBOTS

VIRTUAL ASSISTANTS

SYNTHETIC MEDIA

SELF-DRIVING VEHICLES

FACIAL RECOGNITION

STOCK TRADING

DETECTING THREATS

AUTONOMOUS WEAPONS

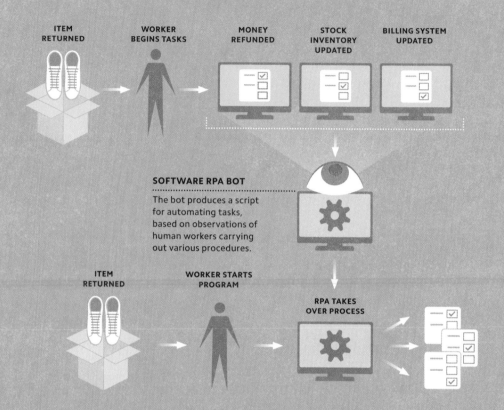

ITEM RETURNED

WORKER BEGINS TASKS

MONEY REFUNDED

STOCK INVENTORY UPDATED

BILLING SYSTEM UPDATED

SOFTWARE RPA BOT

The bot produces a script for automating tasks, based on observations of human workers carrying out various procedures.

ITEM RETURNED

WORKER STARTS PROGRAM

RPA TAKES OVER PROCESS

OFFICE BOTS

The mechanization of repetitive computer-based tasks, such as moving files and entering data, is known as robotic process automation (RPA). Software "bots" mimic human workers' interactions with various computer systems, using a script generated from observations of the procedures. Bots carry out these jobs faster and more reliably than people and do not need a break. Automating mundane activities also frees up human workers so that they can concentrate on more complex tasks.

TRADITIONAL CODING

A software engineer writes all the code for an app.

LOW CODE

Users write some code, but also use visual tools (such as drag and drop menus).

NO CODE

Users build simple apps relying entirely on visual tools, without writing any code.

ACCESSIBLE APP CREATION

Low-code and no-code development platforms enable people without extensive computer coding experience to create apps. Each method provides intuitive visual tools for generating code. Low code simplifies development, although users still require basic coding skills; with no code, however, users do not have to write any code at all. These approaches may help meet the demand for developers. One limitation, however, is that they may be too inflexible to perform many complex tasks. Consequently, these platforms will not replace traditional app development.

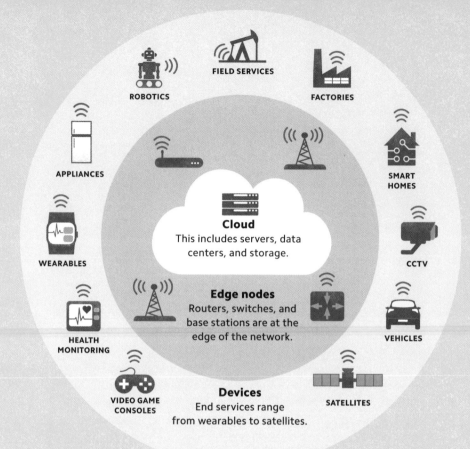

COMPUTING EVERYWHERE

Edge computing involves storing and processing data near the site at which it is generated—either on the device itself or in a local "edge node." This frees up bandwidth by limiting the amount of data traveling to and from major data centers. It also allows data to be processed at much higher speeds and volumes than before, reducing latency (delays in data transfer). This is critical for many Internet of Things (see p.96) applications. For instance, self-driving cars must be able to respond as quickly as possible to their surroundings, as any delay increases the risk of a dangerous accident.

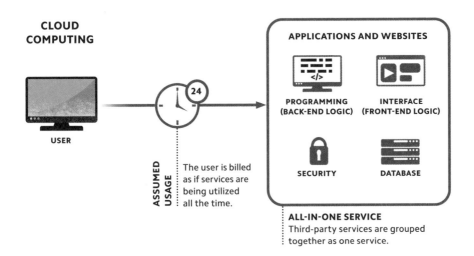

CLOUD COMPUTING

USER

ASSUMED USAGE

The user is billed as if services are being utilized all the time.

APPLICATIONS AND WEBSITES

PROGRAMMING (BACK-END LOGIC)

INTERFACE (FRONT-END LOGIC)

SECURITY

DATABASE

ALL-IN-ONE SERVICE
Third-party services are grouped together as one service.

PAY-AS-YOU-GO SERVERS

In cloud computing, users access remote resources on a server owned by a company, the cloud service provider. These include the code that enables a program or application to operate (back-end logic), which works in tandem with the visual interface on the user's computer screen (front-end logic). In serverless computing, however, the provider allocates services on demand. Instead of paying a fixed fee for a set of resources, regardless of how much they are utilized, the user pays only for actual usage—enabling more efficient handling of limited computing resources.

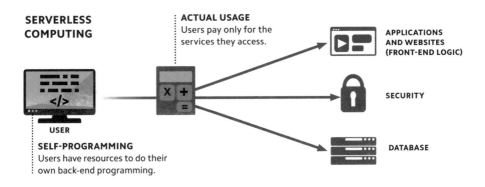

SERVERLESS COMPUTING

USER

ACTUAL USAGE
Users pay only for the services they access.

APPLICATIONS AND WEBSITES (FRONT-END LOGIC)

SECURITY

DATABASE

SELF-PROGRAMMING
Users have resources to do their own back-end programming.

COMPUTING WITH LIGHT

Optical computing uses light to perform operations. Many scientists consider this a promising alternative to conventional computing, mostly because using photons instead of electrons to transmit data could enable a much higher bandwidth: multiple streams of data can be processed simultaneously, using different frequencies of light. Optical computers could theoretically run up to a million times faster than their electronic counterparts. However, it is not yet clear whether optical computing can compete with conventional technology for practical uses.

FROM MECHANICAL TO OPTICAL COMPUTING

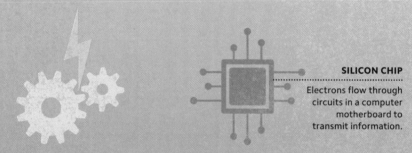

SILICON CHIP
Electrons flow through circuits in a computer motherboard to transmit information.

Electromechanical computing
The first electrically powered computers did not use transistors and other hardware that is familiar today. Instead, they were built from switches, wheels, and relays. They were used in the 1940s and 1950s.

Electronic computing
Electronic computers are the dominant technology used today. To process information, they carry out calculations using electrons in the form of an electrical current flowing through circuits.

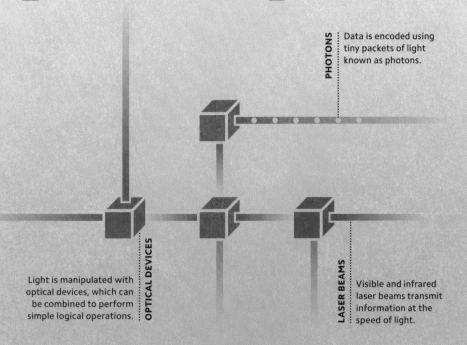

PHOTONS
Data is encoded using tiny packets of light known as photons.

OPTICAL DEVICES
Light is manipulated with optical devices, which can be combined to perform simple logical operations.

LASER BEAMS
Visible and infrared laser beams transmit information at the speed of light.

Optical computing

Optical computers are built from devices that manipulate light to carry out digital computations. They use photons rather than electrons, which enables a higher bandwidth and lower latencies (data delay).

BEYOND THE HARD DRIVE

Every year, tens of zettabytes (1 ZB = 1,000,000,000,000,000,000,000
bytes) of data are generated, a figure that is expected to rise. These
mounting quantities of information must be stored in a manner that is
space- and energy-efficient, secure, easy to access, and stable over
time. Current data storage media, such as hard disk drives,

Data storage
Hard disk drives are the dominant data
storage medium, but systems enabling
greater and more efficient storage are in
development.

Blockchain storage
In blockchain storage (see p.102), data
is saved in linked, encrypted chunks
("blocks"). They are stored on unused
hard drive space in a decentralized
computer network. This can be more
secure than a centralized data center
under a single technology company.

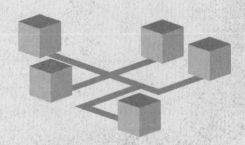

Multicloud storage
Data storage can be
shared across multiple
cloud providers. This
gives the user more
flexibility and security,
while limiting the risks
of being tied to
a single provider.

Crystal storage
Hard drives can
potentially fail after
just a few years, but
storing data in quartz
could keep it stable
for billions of years.
Data is etched onto
a small crystal disk
with a laser.

store "bits"—the smallest piece of information that a computer can process—on tiny areas of a spinning disk. Such media could prove insufficient in the years ahead, however. Meeting this need requires innovation in data storage.

Helium hard drive

Replacing the air in a hard drive with helium can boost performance. As helium is far less dense than air, this minimizes drag and turbulence on the spinning disk, enabling more storage.

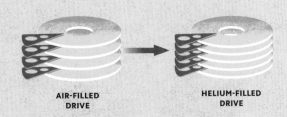

AIR-FILLED DRIVE

HELIUM-FILLED DRIVE

SMR disk drive

Conventional hard drives record data in non-overlapping tracks, but shingled magnetic recording (SMR) drives allow new tracks to overlap slightly with previous tracks, increasing capacity.

CONVENTIONAL DISK DRIVE

SMR DISK DRIVE

DNA storage

Data can be encoded into synthetic strands of DNA and then decoded. This involves converting data into a sequence of the four DNA bases (A, C, G, and T), synthesizing DNA with that sequence, and keeping it in storage, enabling dense data accumulation.

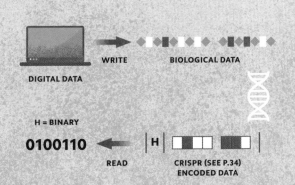

DIGITAL DATA

WRITE

BIOLOGICAL DATA

H = BINARY

0100110

READ

H

CRISPR (SEE P.34) ENCODED DATA

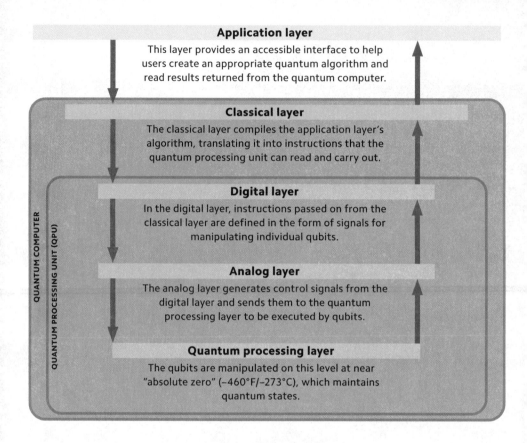

Application layer
This layer provides an accessible interface to help users create an appropriate quantum algorithm and read results returned from the quantum computer.

Classical layer
The classical layer compiles the application layer's algorithm, translating it into instructions that the quantum processing unit can read and carry out.

Digital layer
In the digital layer, instructions passed on from the classical layer are defined in the form of signals for manipulating individual qubits.

Analog layer
The analog layer generates control signals from the digital layer and sends them to the quantum processing layer to be executed by qubits.

Quantum processing layer
The qubits are manipulated on this level at near "absolute zero" (–460°F/–273°C), which maintains quantum states.

QUANTUM COMPUTER

QUANTUM PROCESSING UNIT (QPU)

SOLVING THE IMPOSSIBLE

A quantum computer utilizes the way matter behaves at a tiny or "quantum" scale, working with atoms or subatomic particles (such as protons or electrons) to solve problems. This gives them the potential to perform tasks that are almost impossible for classical computers (which manipulate data in the form of "bits," represented by either "0" or "1"), from simulating complex physical systems to cracking encryption protocols. However, building and running quantum computers remains a vast engineering challenge, and classical computers still outperform quantum computers for practical uses.

EVERYWHERE AT ONCE

In classical computing, the basic unit of information is a bit, a binary digit that can exist in one of two states, represented by "0" or "1." In quantum computing, the basic unit is a quantum bit or "qubit," a subatomic particle. Qubits can exist in two possible states at once ("superposition"). During measurement (the manipulation of qubits to obtain a numerical result), this superposition collapses, leaving the qubit in only one state. In theory, superposition and entanglement (see below) make quantum computers more powerful than classical computers.

Bits

Classical computing is based on the manipulation of bits. A bit can exist in one state or another, represented by either "0" or "1."

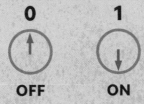

0
OFF

1
ON

Qubits

Quantum computing is based on the manipulation of qubits. Unlike a bit, a qubit can exist not only in one of two states but also in a combination of both states simultaneously.

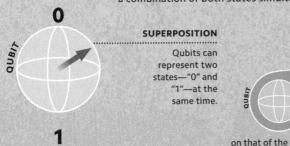

0

QUBIT

1

SUPERPOSITION

Qubits can represent two states—"0" and "1"—at the same time.

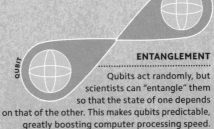

QUBIT

QUBIT

ENTANGLEMENT

Qubits act randomly, but scientists can "entangle" them so that the state of one depends on that of the other. This makes qubits predictable, greatly boosting computer processing speed.

QUANTUM IN PRACTICE

Despite progress in the development of delicate hardware for quantum computers, they currently have almost no practical applications. Much related research is focused on finding uses for them. They could be invaluable in solving problems that have vast numbers of options: classical computers must assess each possibility in turn, but quantum computers can consider many possibilities at once. This could transform cryptography, which is largely based on mathematical problems that are impractical for classical computers to solve.

Looking for work
Aside from fields such as combinatorics and random number generation, there are few practical applications for quantum computing at present.

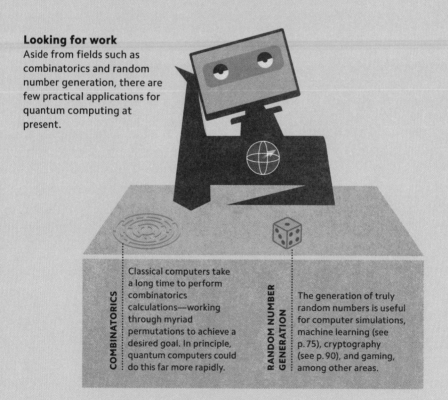

COMBINATORICS
Classical computers take a long time to perform combinatorics calculations—working through myriad permutations to achieve a desired goal. In principle, quantum computers could do this far more rapidly.

RANDOM NUMBER GENERATION
The generation of truly random numbers is useful for computer simulations, machine learning (see p. 75), cryptography (see p. 90), and gaming, among other areas.

CLASSICAL VS. QUANTUM

Quantum supremacy is a symbolic goal of quantum computing:
to demonstrate that a quantum computer is capable of solving
problems that would be essentially impossible for classical computers.
Reaching quantum supremacy has two major complications. The first
is finding a problem for which quantum computers have a definitive
advantage. The second is building a high-qubit quantum computer
while managing decoherence—the interaction of qubits (see p.87)
with their environment, causing disturbance and loss of information.

Impossibly big problems?
As certain problems grow, the time
and resources that classical computers
need to solve them increase
exponentially. In theory, quantum
computers are far more efficient.

SUPER-FAST SOLUTIONS
A quantum computer could solve in
a few minutes some problems that
would occupy a classical computer
for thousands of years.

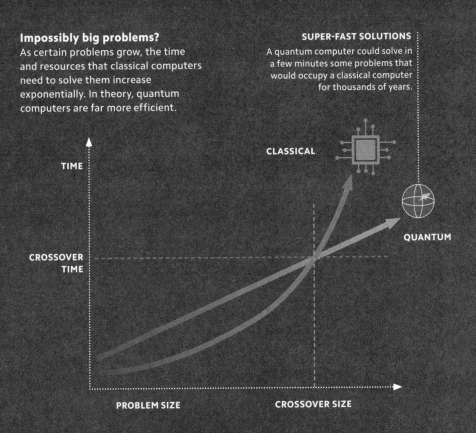

TIME

CLASSICAL

QUANTUM

CROSSOVER
TIME

PROBLEM SIZE

CROSSOVER SIZE

Filtering streamed protons

The sender uses four filters, which are randomly swapped to generate a stream of polarized photons (light particles), each of which represents "0" or "1."

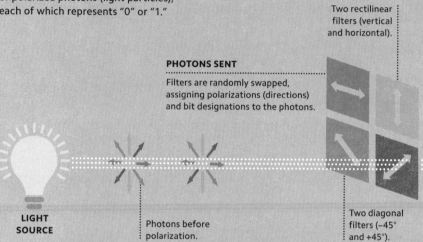

Two rectilinear filters (vertical and horizontal).

PHOTONS SENT

Filters are randomly swapped, assigning polarizations (directions) and bit designations to the photons.

LIGHT SOURCE

Photons before polarization.

Two diagonal filters (−45° and +45°).

UNBREAKABLE CODES

Cryptography involves encrypting data so that only someone with the correct "key" can decrypt it. Quantum cryptography is considered safer than classical cryptography, as it relies on the laws of physics—rather than a complex mathematical problem—for security. The best-known example is quantum key distribution, in which a sender and receiver exchange particles in quantum states (representing bits) to generate a secret key. A quantum state cannot be measured without being disturbed, so an "eavesdropper" (hacker) cannot intercept the particles without being detected.

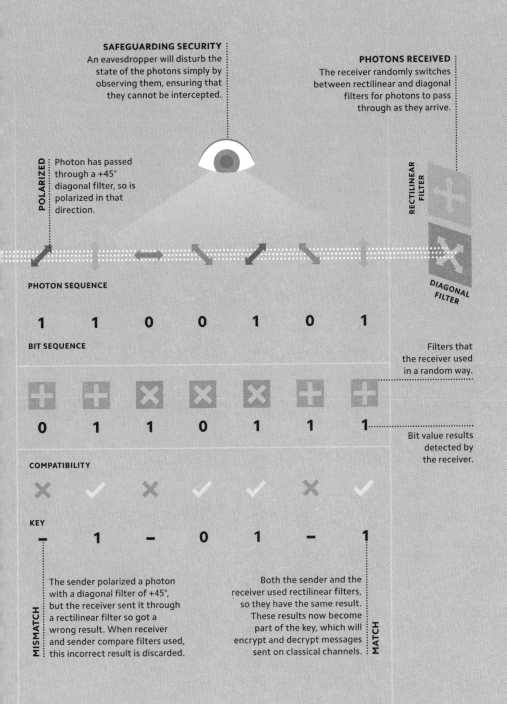

SAFEGUARDING SECURITY
An eavesdropper will disturb the state of the photons simply by observing them, ensuring that they cannot be intercepted.

PHOTONS RECEIVED
The receiver randomly switches between rectilinear and diagonal filters for photons to pass through as they arrive.

POLARIZED
Photon has passed through a +45° diagonal filter, so is polarized in that direction.

RECTILINEAR FILTER

DIAGONAL FILTER

PHOTON SEQUENCE

1   1   0   0   1   0   1

BIT SEQUENCE

Filters that the receiver used in a random way.

0   1   1   0   1   1   1

Bit value results detected by the receiver.

COMPATIBILITY

KEY

–   1   –   0   1   –   1

MISMATCH
The sender polarized a photon with a diagonal filter of +45°, but the receiver sent it through a rectilinear filter so got a wrong result. When receiver and sender compare filters used, this incorrect result is discarded.

MATCH
Both the sender and the receiver used rectilinear filters, so they have the same result. These results now become part of the key, which will encrypt and decrypt messages sent on classical channels.

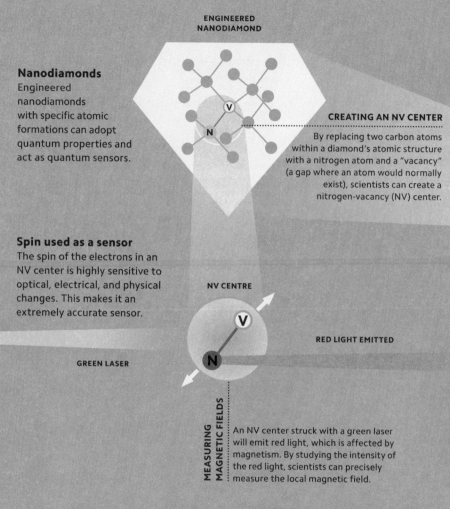

ENGINEERED NANODIAMOND

Nanodiamonds
Engineered nanodiamonds with specific atomic formations can adopt quantum properties and act as quantum sensors.

CREATING AN NV CENTER
By replacing two carbon atoms within a diamond's atomic structure with a nitrogen atom and a "vacancy" (a gap where an atom would normally exist), scientists can create a nitrogen-vacancy (NV) center.

Spin used as a sensor
The spin of the electrons in an NV center is highly sensitive to optical, electrical, and physical changes. This makes it an extremely accurate sensor.

NV CENTRE

RED LIGHT EMITTED

GREEN LASER

MEASURING MAGNETIC FIELDS

An NV center struck with a green laser will emit red light, which is affected by magnetism. By studying the intensity of the red light, scientists can precisely measure the local magnetic field.

DELICATE DETECTORS

Although the sensitivity of quantum states presents challenges when it comes to building quantum computers, it is ideal for taking measurements. Quantum sensors, such as engineered NV centers within nanodiamonds, exploit quantum phenomena to measure

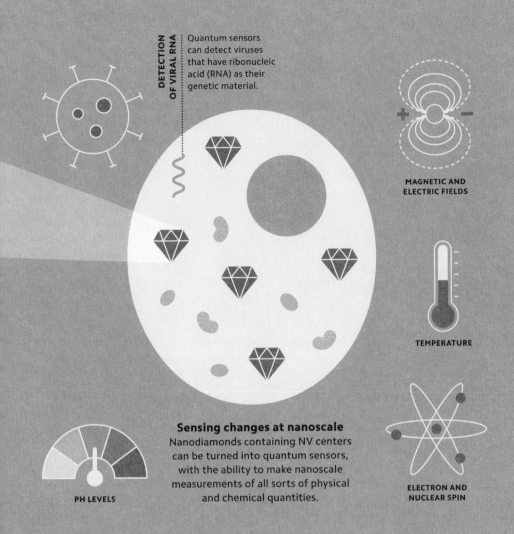

DETECTION OF VIRAL RNA
Quantum sensors can detect viruses that have ribonucleic acid (RNA) as their genetic material.

MAGNETIC AND ELECTRIC FIELDS

TEMPERATURE

Sensing changes at nanoscale
Nanodiamonds containing NV centers can be turned into quantum sensors, with the ability to make nanoscale measurements of all sorts of physical and chemical quantities.

PH LEVELS

ELECTRON AND NUCLEAR SPIN

variables—including electric and magnetic fields, and PH levels—far more precisely than classical devices can. Current research using quantum sensors includes tracking neural activity by measuring the tiny magnetic fields that arise from electrical flow in the brain.

COMMUN AND MEDIA

ICATIONS

Today tens of billions of devices speak to each other through ever-expanding communications infrastructure, such as 5G and satellite constellations. This is known as the Internet of Things (IoT). The "original" internet is entering a new phase that may see it become more decentralized and democratic, with technologies like blockchain used to distribute control. The barrier between body and computer is being eroded, with more digital experiences beyond the screen of a phone or computer. Devices such as VR headsets immerse users in virtual worlds, with input from the body. Some people hope that it might even be possible to live in computers within their lifetimes.

CONNECTED LIVING

The Internet of Things (IoT) refers to the ever-growing network of billions of objects incorporating sensors, actuators, and communication devices. This enables more and more of the physical world to be precisely monitored and controlled. For example, "smart homes" may include an air-quality monitor that provides detailed insights and recommendations, or an oven that the owner can start heating up remotely using an app. The IoT also encompasses smart cities (see p.142) and precision farming (see p.47).

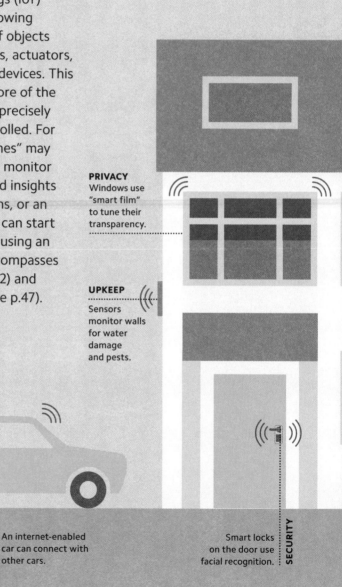

PRIVACY
Windows use "smart film" to tune their transparency.

UPKEEP
Sensors monitor walls for water damage and pests.

ONLINE VEHICLE
An internet-enabled car can connect with other cars.

SECURITY
Smart locks on the door use facial recognition.

> **"The Internet of Things is about creating a smart, more connected world."**
> Eric Schmidt, entrepreneur

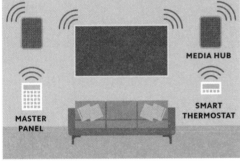

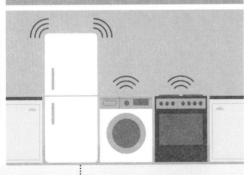

OVERSEEING DEVICE USAGE

Smart lighting turns itself on and off as required, while smart plugs control the flow of electricity to all devices.

MEDIA HUB

MASTER PANEL

SMART THERMOSTAT

REMOTE CONTROL

Smart home devices can be checked and given commands from their companion apps. This makes them accessible from anywhere in the world.

SMART KITCHEN

Many kitchen appliances can be controlled remotely. Some may have AI capabilities—for instance, a fridge that "sees" its contents and suggests recipes.

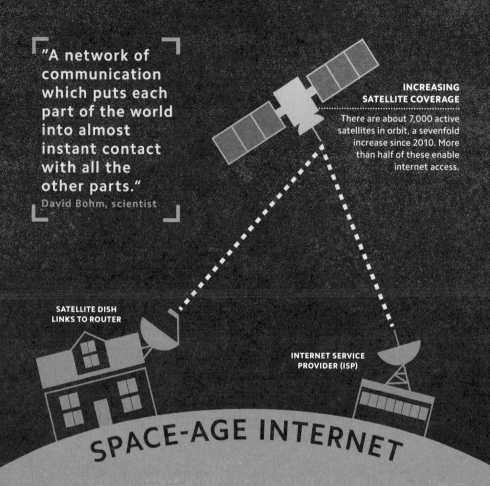

"A network of communication which puts each part of the world into almost instant contact with all the other parts."
David Bohm, scientist

INCREASING SATELLITE COVERAGE

There are about 7,000 active satellites in orbit, a sevenfold increase since 2010. More than half of these enable internet access.

SATELLITE DISH LINKS TO ROUTER

INTERNET SERVICE PROVIDER (ISP)

SPACE-AGE INTERNET

A satellite can establish a communication channel between two distant locations on Earth, relaying and amplifying the radio signal between transmitter and receiver. This enables wireless communication that would otherwise be obstructed by Earth's curvature. Although communications satellites have existed for decades, it is only recently that a large number of satellites has been available to provide these links. Satellite internet access, increasingly supported by satellite constellations (see opposite), is useful for people in rural areas with limited—or no—ground-based internet infrastructure.

GLOBAL COVERAGE

A satellite constellation is a group of satellites working as a unified system. One satellite has limited coverage, but a constellation provides complete global coverage. Perhaps the most famous constellation is the global positioning system (GPS), an aid to navigation that pinpoints locations on Earth using more than 30 satellites. Some have many more, though: the Starlink constellation, which provides internet coverage, contains thousands of satellites. The falling cost of launches, thanks largely to reusable rockets (see p.69), has made it possible to produce more satellite constellation projects.

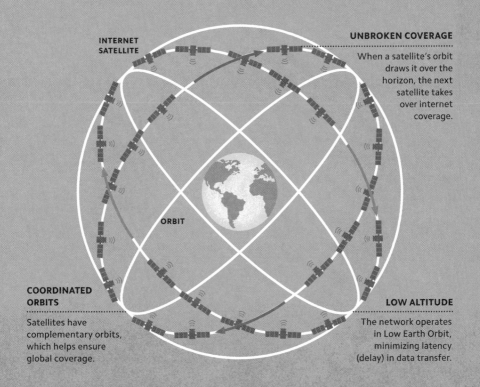

INTERNET SATELLITE

UNBROKEN COVERAGE

When a satellite's orbit draws it over the horizon, the next satellite takes over internet coverage.

ORBIT

COORDINATED ORBITS

Satellites have complementary orbits, which helps ensure global coverage.

LOW ALTITUDE

The network operates in Low Earth Orbit, minimizing latency (delay) in data transfer.

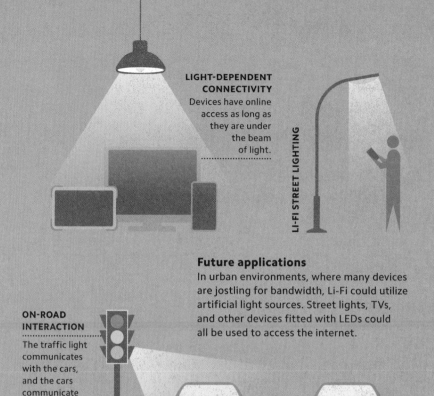

LIGHT-DEPENDENT CONNECTIVITY
Devices have online access as long as they are under the beam of light.

LI-FI STREET LIGHTING

Future applications
In urban environments, where many devices are jostling for bandwidth, Li-Fi could utilize artificial light sources. Street lights, TVs, and other devices fitted with LEDs could all be used to access the internet.

ON-ROAD INTERACTION
The traffic light communicates with the cars, and the cars communicate with each other.

SPEAKING WITH LIGHT

While Wi-Fi employs routers and radio waves to communicate wirelessly, Li-Fi uses LED lights and light waves. LEDs stream data in the form of light signals flickering at speeds that are imperceptible to the human eye. The light is converted back into electronic data on a user's device. Li-Fi enables faster data transfer than Wi-Fi and has a greater bandwidth, reaching places Wi-Fi cannot penetrate (such as underwater). Li-Fi is not yet in widespread use, but there is interest in the technology.

FUTURE NET?

The internet has had two major phases: Web 1.0, in which users consumed content on static web pages, and Web 2.0, in which users create and consume content on centralized platforms. Web 3.0 is characterized as more decentralized, with widespread use of blockchain (see p.102) and AI (see pp.76–77). However, there is dispute over how Web 3.0 should be defined, and whether any of those definitions describe the direction in which the internet is moving. For instance, rather than diversifying, the internet appears to be becoming more centralized and dominated by a small number of tech companies.

Where could 3.0 go?
Despite conflicting definitions, key features are associated with Web 3.0. These generally relate to providing individual users with greater control.

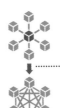

DECENTRALIZATION
Peer-to-peer networks, such as blockchain, avoid being overseen by a central entity.

CONNECTIVITY
Information is accessible from a huge range of connected devices (see p.96).

SEMANTIC WEB
Machines are enabled to "understand" information on the internet.

3D GRAPHICS
Three-dimensional visuals are more pervasive, especially in "virtual worlds" (see p.105).

AI AND MACHINE LEARNING
AI technologies—such as generative AI (see p.104)—become integrated across the internet.

PERMISSIONLESS
Participation is not dependent on authorization from a central authority.

BLOCK BY BLOCK

A distributed ledger is a database that is stored across many computers, which are spread out geographically. The best-known use is blockchain: a decentralized database of records ("blocks") linked via cryptography. Blocks cannot be altered without the consensus of all participants. Blockchain is associated with cryptocurrencies, for which every transaction is recorded on the relevant blockchain. It could also be used in supply-chain management, for example, to ensure that diamonds originate from ethical sources.

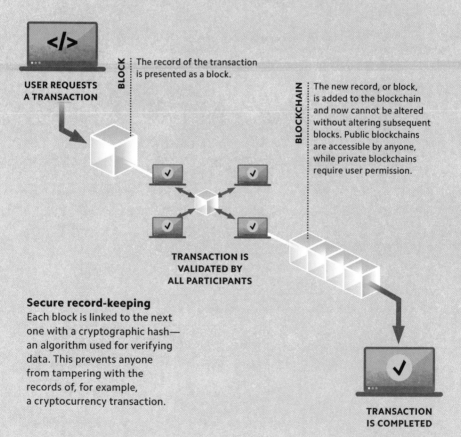

USER REQUESTS A TRANSACTION

BLOCK
The record of the transaction is presented as a block.

BLOCKCHAIN
The new record, or block, is added to the blockchain and now cannot be altered without altering subsequent blocks. Public blockchains are accessible by anyone, while private blockchains require user permission.

TRANSACTION IS VALIDATED BY ALL PARTICIPANTS

Secure record-keeping
Each block is linked to the next one with a cryptographic hash—an algorithm used for verifying data. This prevents anyone from tampering with the records of, for example, a cryptocurrency transaction.

TRANSACTION IS COMPLETED

"Write a 100,000-word period novel set in London. Include a jewel robbery and a master detective."

The user types a prompt or question for the model in natural language.

ENCODER

The AI program's encoder creates an abstract representation of the input that the model can "understand."

DECODER LOOP

The decoder keeps generating words until it produces a signal to stop.

DECODER

The AI program's decoder uses this representation to generate a sequence of text, word by word.

OUTPUT

The model responds to the user in natural language.

"Smog unfurled in the glow of the gas lamp that stood outside the jewelry shop…"

OUTPUT

TALKING TO MACHINES

The ability of a machine to understand human language is known as natural language processing (NLP). This field of AI has undergone considerable advances in recent years, with troves of language data harvested from the internet being used to train models (see p.76) to read, write, listen, and speak. One example is large language models, which are a type of generative AI (see p.104) that produces text indistinguishable from that written by humans. These models work by repeatedly predicting the next word in a sequence.

AI ARTISTS

Generative AI is the field of artificial intelligence dedicated to creating new text, audio, images, and other media. Models are trained using large datasets of existing content. For instance, text-to-image models learn from images taken from the internet and tagged with text descriptions. The field's rapid rise has raised questions, such as whether an AI can "own" content and how to handle convincing disinformation, such as fake photos of real people. In future, encrypted watermarks could identify authentic media.

SOURCING ONLINE
An AI model can "scrape" the internet, collecting data from sources such as stock images and real artworks.

MAKING ART
Based on the collected source material, the AI model generates a new artwork.

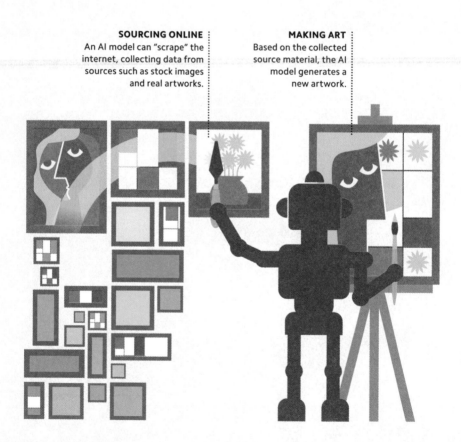

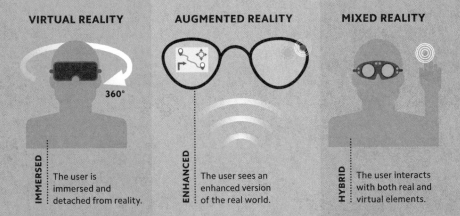

VIRTUAL REALITY

360°

IMMERSED
The user is immersed and detached from reality.

AUGMENTED REALITY

ENHANCED
The user sees an enhanced version of the real world.

MIXED REALITY

HYBRID
The user interacts with both real and virtual elements.

ENHANCED ENVIRONMENTS

Extended reality (XR) describes technologies that enrich or replace a user's environment with digital images. Virtual reality sets users in an artificial world, typically via a wearable display. Augmented reality layers digital elements onto a user's view of the world; mixed reality is a more interactive extension of this. XR can be used by engineers to put themselves and clients "into" a plan, for example, or in health care to create immersive environments to test patients' responses.

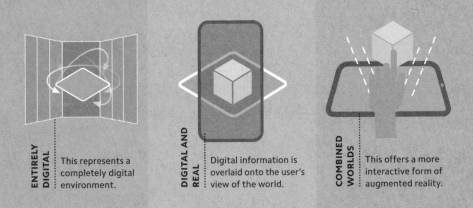

ENTIRELY DIGITAL
This represents a completely digital environment.

DIGITAL AND REAL
Digital information is overlaid onto the user's view of the world.

COMBINED WORLDS
This offers a more interactive form of augmented reality.

Fantasy fitness
Users with virtual reality hardware can work out at home while presenting as an avatar of their choice in a virtual exercise class.

INTO THE METAVERSE...

The metaverse concept, first described in the science-fiction novel *Snow Crash* by Neal Stephenson, is a hypothetical version of the internet. It imagines the internet as a single, shared 3D virtual world in which users are represented by avatars, typically accessed using VR hardware (see opposite, and p.105). Users could work, shop, socialize, and do everything else they do online within this more immersive environment. This vision for the metaverse has been met with little enthusiasm, but metaverse-like games are extremely popular, with hundreds of millions of active users. These games tend to offer users an "escape" from the mundane into fantastical virtual worlds.

AT THE WAVE OF A HAND

Most computers are operated by using mice, keyboards, touch screens, and sometimes microphones. Gesture-based computing allows humans to interact with computers beyond these conventional input devices by incorporating movements of the head and hands, eye movements, posture, and even facial expressions. This range of 3D inputs could make experiences in extended reality (see p.105) more immersive and intuitive. Specialized hardware, such as wired gloves, is often required, although a single webcam can be used to detect some simple gestures. Gesture-based computing is well suited to applications across gaming, smart homes, healthcare, and robotics, and could make it easier for people with certain disabilities to use computers.

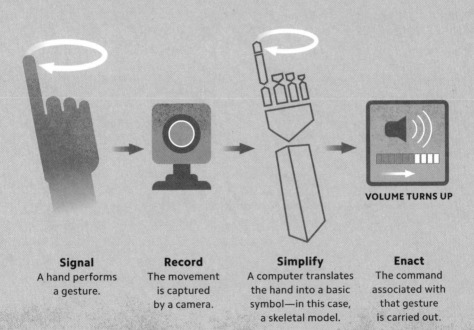

VOLUME TURNS UP

Signal
A hand performs a gesture.

Record
The movement is captured by a camera.

Simplify
A computer translates the hand into a basic symbol—in this case, a skeletal model.

Enact
The command associated with that gesture is carried out.

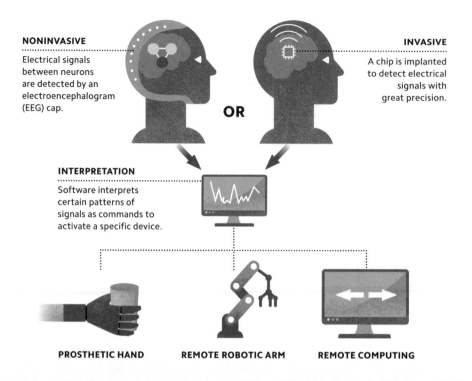

NONINVASIVE

Electrical signals between neurons are detected by an electroencephalogram (EEG) cap.

INVASIVE

A chip is implanted to detect electrical signals with great precision.

OR

INTERPRETATION

Software interprets certain patterns of signals as commands to activate a specific device.

PROSTHETIC HAND

REMOTE ROBOTIC ARM

REMOTE COMPUTING

MIND CONTROL

A brain-computer interface (BCI) is a connection between a brain and an external device, achieved by either an invasive or noninvasive method. Many technologies could be called BCIs, but the term generally refers to a way of enabling humans to control devices—such as a computer or a robotic prosthesis—with their thoughts. BCIs detect and analyze electrical signals that arise in the brain, and convert them into commands for the device. The main application for BCIs is to replace or restore a human function limited by disease or injury, although there is interest in nonmedical applications, including gaming and defense.

DIGITAL IMMORTALITY

To some people, eternal existence in digital form ("digital immortality") may one day provide a feasible means of living forever. The process could involve scanning a brain and storing it in digital form so that an individual's personality, memories, and (according to some schools of thought) conscience continue after bodily death. This presents the possibility of being represented by an avatar, or of controlling a robotic body. The technology is entirely speculative at present, with scientists only able to simulate entire brains of very simple organisms, such as the worm *Caenorhabditis elegans*.

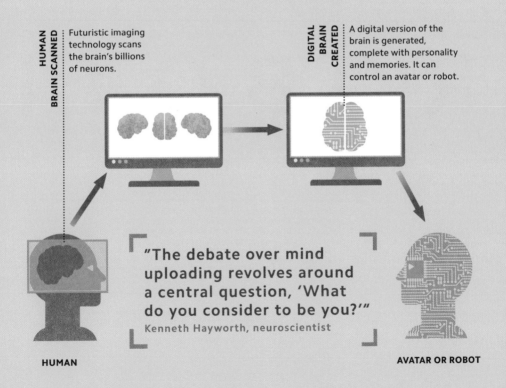

HUMAN BRAIN SCANNED
Futuristic imaging technology scans the brain's billions of neurons.

DIGITAL BRAIN CREATED
A digital version of the brain is generated, complete with personality and memories. It can control an avatar or robot.

HUMAN

AVATAR OR ROBOT

"The debate over mind uploading revolves around a central question, 'What do you consider to be you?'"
Kenneth Hayworth, neuroscientist

ROBOTI

C S

Robots do jobs that are dull, dirty, dangerous, or almost impossible for humans to do. They can work around the clock in factories; search for survivors following natural disasters; or be launched on one-way trips into space. They come in a vast range of sizes, with some of the smallest robots made from biological components such as cells. Robots have varying degrees of autonomy, from closely controlled military drones and surgical machines to AI-powered robots acting without human supervision. A major challenge is creating robots that work well alongside people. Approaches can involve designing them with soft materials, humanoid characteristics, and artificial social intelligence.

REMOTE SURGERY

Specialized robotic systems can be used to carry out surgical procedures with enhanced precision and control. The most common system, the da Vinci system, is used for minimally invasive (keyhole) surgery. Instead of using handheld instruments, the surgeon sits at a console and guides the instruments remotely. This system reduces fatigue, eradicates tremor, and allows surgeons to work from remote locations. Technologies in development today could mean that future robotic surgeons have greater autonomy and practice more complex surgery.

DETAILED VIEW

The surgeon receives a 3D image of the site to be operated on.

IN CONTROL

The surgeon uses a console to manipulate the system's instruments.

From a distance
Robotic surgery could be performed by a surgeon in a different continent to the patient.

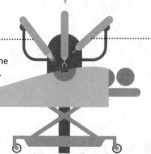

ALL-SEEING
A camera streams the operation to the surgeon's monitor.

ROCK STEADY
Robotic arms have a wrist function at their tips, allowing them to turn the surgical tools that they hold.

ROBOTS IN THE FACTORY

There are millions of industrial robots at work today. Most of them are installed on assembly lines, where they each carry out a task over and over with great consistency. The most common example of this technology is a "robotic arm." This approximately resembles a human arm and has a useful attachment on the end, such as a gripper or electrode, and several degrees of freedom that allow it a wide range of movement. Industrial robots are well suited for smart factories—manufacturing facilities in which data from connected devices is used to optimize processes.

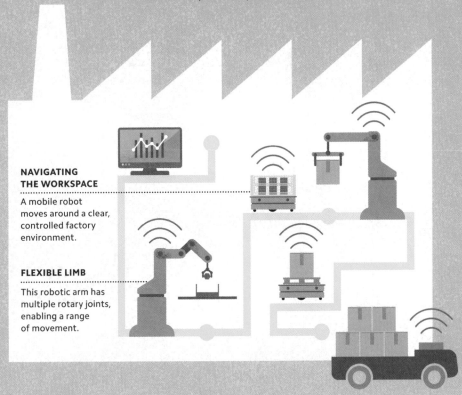

NAVIGATING THE WORKSPACE

A mobile robot moves around a clear, controlled factory environment.

FLEXIBLE LIMB

This robotic arm has multiple rotary joints, enabling a range of movement.

ROBOTS TO
THE RESCUE

Robots are becoming invaluable life-saving tools in disaster zones. A huge variety of robots have been designed to assist rescue operations, with tasks including searching for survivors, moving rubble, and delivering medical supplies. Often inspired by animal behavior, these robots fly, swim, trundle over debris, and wriggle through small gaps to perform tasks that might be too dangerous for humans.

DRONE
A drone with an arm can reach, and search, otherwise inaccessible places.

Crew of robots
A team of distinct search and rescue robots can be deployed at the scene of a disaster, each with its own task.

Hydrogel has long, crosslinked molecules.

GRIPPER PRINTED FROM HYDROGEL

Hydrogel gripper
A clasping device printed from hydrogel can catch a fish without harming it.

The robotic gripper is made from hydrogel structures connected to rubbery tubes.

OPEN GRIPPER

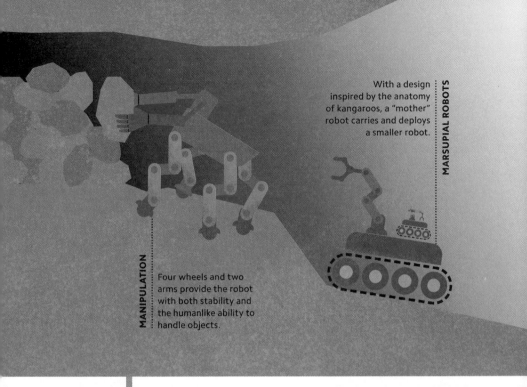

With a design inspired by the anatomy of kangaroos, a "mother" robot carries and deploys a smaller robot.

Four wheels and two arms provide the robot with both stability and the humanlike ability to handle objects.

Hydrogel structure holds water molecules.

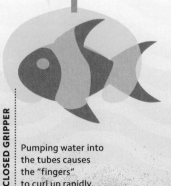

CLOSED GRIPPER

Pumping water into the tubes causes the "fingers" to curl up rapidly.

A HUMAN TOUCH

Robots are typically built from rigid materials like metals and hard plastics. The field of soft robotics focuses on creating robots from hydrogels, silicone, rubber, and other materials that resemble living tissues. These are useful when the robot needs to adapt its shape to survive an impact, or to handle an object with the delicacy that a human would apply—for instance, helping health-care workers handle patients.

Microswimmer
A soft microbot, inspired by a parasite that swims through bloodstreams, could unclog arteries.

Robotic crab
A 0.02 in (0.5 mm) wide "crawling" robot moves by changing shape as it heats and cools.

Self-folding robot
A nanoscale device can fold itself into a 3D shape with a jolt of electricity, enabling it to move.

MINIATURE ROBOTS

Microbots (robots on the microscale) and nanobots (robots on the nanoscale) present a particular set of opportunities and challenges. They have the potential to be especially useful in medicine by entering the human body to perform delicate procedures, such as breaking up blood clots. Microbots and, particularly, nanobots remain mostly experimental, not least because engineers are still figuring out how to power them at such a small scale. A possible solution to this problem may be to build them with biological components—for instance, incorporating a bacterium or sperm cell for propulsion.

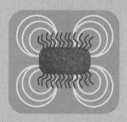

Long-haired robot
The hairs of a larva-inspired robot oscillate in response to sound, allowing it to swim.

Natural motor
Kinesin, a protein, "walks" along cellular structures. It could be used as a motor for nanobots.

Nanocar
A nanoscale "car" rolls in response to temperature, allowing it to travel inside the human body.

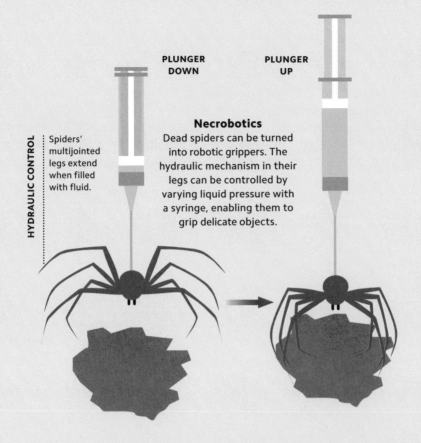

HYDRAULIC CONTROL

Spiders'
multijointed
legs extend
when filled
with fluid.

Necrobotics
Dead spiders can be turned
into robotic grippers. The
hydraulic mechanism in their
legs can be controlled by
varying liquid pressure with
a syringe, enabling them to
grip delicate objects.

WORKING WITH NATURE

The field of biorobotics involves both robotics inspired by
biological systems and robotics integrated with biological systems.
Cells, tissues, and entire organisms are used to build robots, utilizing
nature's complex machinery. Skeletal muscle from mammals and
dorsal vessel (circulatory) tissue from insects, for example, have
been used as actuators, creating movement, and dead spiders
have been repurposed as robotic grippers using necrobotics
(see above). As well as being biodegradable, these
robots are self-powered and self-healing, if living.

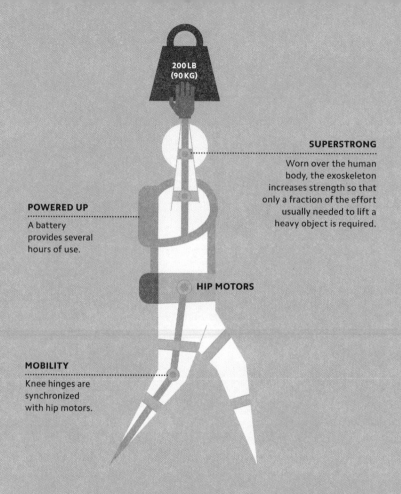

200LB
(90KG)

SUPERSTRONG

Worn over the human body, the exoskeleton increases strength so that only a fraction of the effort usually needed to lift a heavy object is required.

POWERED UP

A battery provides several hours of use.

HIP MOTORS

MOBILITY

Knee hinges are synchronized with hip motors.

SUPERHUMAN STRENGTH

Powered exoskeletons are large devices that enhance the wearer's physical abilities. Sensors throughout the structure relay data to a central control system, which then directs signals to actuators that synchronize the exoskeleton with the user's intended movements. Exoskeletons are being used to help patients recover functions (especially lower limb function) following injury or illness. However, there is also interest in industrial and military applications, with the aim of endowing the wearer with superhuman strength.

HIVE MINDS

Modeled on the cooperative actions
of insects such as ants, swarm robotics
creates intelligent collective behavior in a
group of simple robots. Intelligent behavior
is learned from interactions between the
robots, and between the robots and their
surroundings. The swarm is fault tolerant,
continuing to work when some members of
the group fail. Swarm robotics could be used
in surveillance, construction, and nature
restoration. Bee-inspired robots, for
example, could pollinate plants.

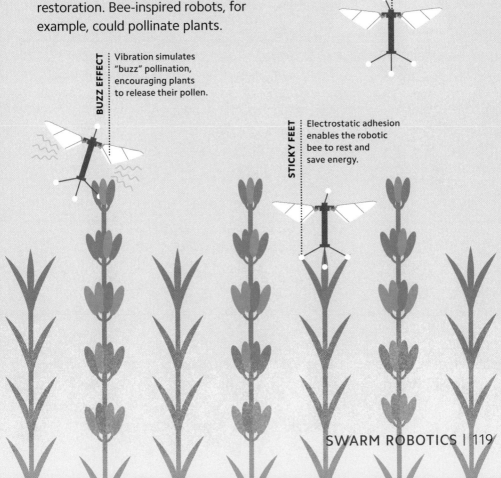

ENERGY
Solar panels
power the
robotic bees.

BUZZ EFFECT
Vibration simulates
"buzz" pollination,
encouraging plants
to release their pollen.

STICKY FEET
Electrostatic adhesion
enables the robotic
bee to rest and
save energy.

BRAIN AND BODY

Embodying AI brings digital AI into the physical world. This means giving it a physical "body" that can sense and interact with its surroundings. These AI-controlled robots are equipped with a multitude of sensors and actuators—devices that translate signals into movement. The AI may learn from its interactions with the physical world—for example, robotic vacuum cleaners will gradually map the layout of a home. Embodied AI may take humanoid form (see opposite) if this is relevant for its work, for instance, being a helper or companion for a human.

LIVING WITH ROBOTS

ENHANCED COMMUNICATION

The robot mimics human interaction by turning its head to make eye contact and speaking.

3D MAPPING

Cameras and other sensors enable the robot to detect its surroundings and map out its route.

ATTACHED TRAY

Tray is preloaded with food and drink.

SELF-DRIVING TECH

To navigate around obstacles, robots use self-driving technology (see p.63).

Today, most robots operate in constrained environments, such as warehouses and factories. Unpredictable "real-world" environments, such as restaurants (see above), hospitals, or transportation hubs, are more challenging for them. They must be able to navigate complex and changeable surroundings, possibly climbing stairs or operating door handles. In some contexts, they must also be able to interact with humans using speech (see p.103) and by interpreting gestures and facial expressions.

KILLER ROBOTS

Defense has long been a driving force behind robotics research and development. Using robots to carry out dangerous military operations that would ordinarily have been performed by people can reduce the human cost of war. Aerial drones—uncrewed flying vehicles—are widely deployed. They are used for surveillance and strikes and have varying degrees of autonomy, although they remain mostly under human control. Other military robots include land and sea drones, exoskeletons (see p.118), uncrewed vehicles for collecting casualties, and machines for laying and clearing mines. More controversial are lethal autonomous weapons that are capable of locking onto and firing at targets.

ROBOT EXPLORERS

Space agencies have been building robots since the 1960s, making space exploration possible without the costs and complications of using astronauts. Some space robots operate independently, such as NASA's Mars rovers, which trundle over the surface of the planet studying its rocks and sending data back to Earth. Others, such as the humanoid "robonaut" on the International Space Station, assist astronauts. New robots are being developed all the time to carry out more ambitious space missions. Some commercial companies are planning to use robots to mine asteroids for valuable metals.

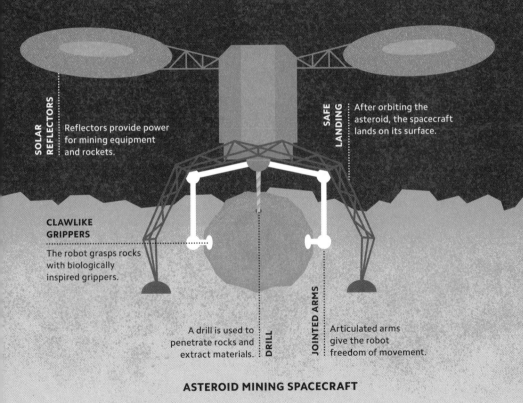

SOLAR REFLECTORS

Reflectors provide power for mining equipment and rockets.

SAFE LANDING

After orbiting the asteroid, the spacecraft lands on its surface.

CLAWLIKE GRIPPERS

The robot grasps rocks with biologically inspired grippers.

DRILL

A drill is used to penetrate rocks and extract materials.

JOINTED ARMS

Articulated arms give the robot freedom of movement.

ASTEROID MINING SPACECRAFT

ENERGY

The energy sector is undergoing rapid transformation. Finding alternative energy sources to fossil fuels (coal, oil, and gas) is essential to avert catastrophic global warming, reduce energy insecurity, and minimize waste. This requires a huge expansion in the use of renewable energy, including developing new technologies, improving carbon capture, and optimizing the distribution and storage of energy. The development of smart power grids, giant batteries, and bladeless turbines are just some of the innovations that could revolutionize the sector.

FUTURE GRID

Conventional power grids distribute electricity from large power stations fueled by gas, coal, or nuclear to points of consumption. Renewable energy sources, such as wind and solar, which fluctuate with the weather, need to be managed by a "smart grid" that takes such variables into account. In a smart grid, diverse energy sources flow to the end user, and the end user returns excess energy, for example, from rooftop solar. Digital technology manages this complex relationship of supply and demand. Smart grids also use green energy efficiently. For instance, electricity generated by wind turbines at night, when demand is low, can charge up grid-scale batteries (see p.132) for use when demand is high.

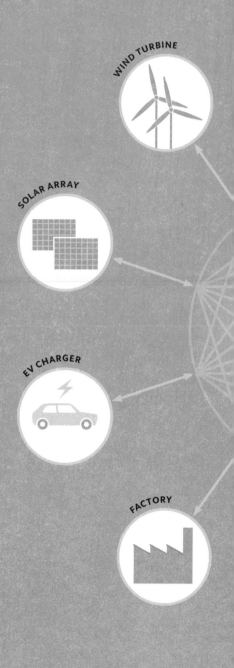

WIND TURBINE

SOLAR ARRAY

EV CHARGER

FACTORY

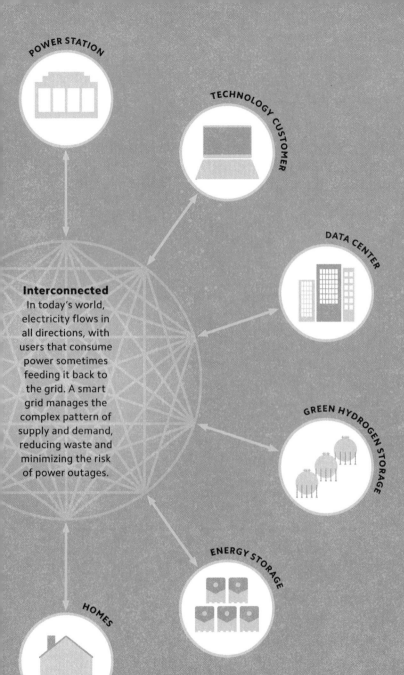

POWER STATION

TECHNOLOGY CUSTOMER

DATA CENTER

GREEN HYDROGEN STORAGE

ENERGY STORAGE

HOMES

Interconnected
In today's world, electricity flows in all directions, with users that consume power sometimes feeding it back to the grid. A smart grid manages the complex pattern of supply and demand, reducing waste and minimizing the risk of power outages.

FROM GRAY TO GREEN

Hydrogen, the most abundant element in the universe, is a cleaner alternative to natural gas, a fossil fuel. Hydrogen can be used for heating, electricity, industrial processes (including in hard-to-decarbonize sectors such as steelmaking), and to power vehicles in hydrogen fuel cells (see p.61). While hydrogen itself does not generate harmful carbon dioxide, its production usually does. This is known as "gray hydrogen." Combining this process with carbon capture and

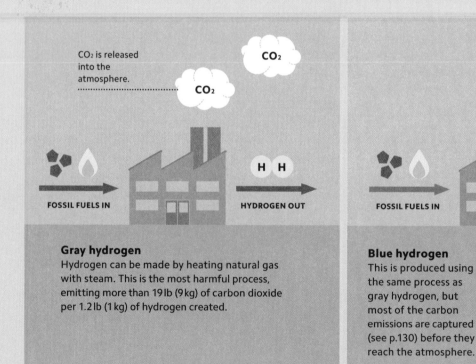

CO_2 is released into the atmosphere.

CO_2

CO_2

FOSSIL FUELS IN

H H

HYDROGEN OUT

FOSSIL FUELS IN

Gray hydrogen
Hydrogen can be made by heating natural gas with steam. This is the most harmful process, emitting more than 19 lb (9 kg) of carbon dioxide per 1.2 lb (1 kg) of hydrogen created.

Blue hydrogen
This is produced using the same process as gray hydrogen, but most of the carbon emissions are captured (see p.130) before they reach the atmosphere.

storage, known as blue hydrogen, is a cleaner alternative. Hydrogen can also be created via electrolysis. Known as green hydrogen, this expensive process accounts for less than 0.1 percent of all hydrogen.

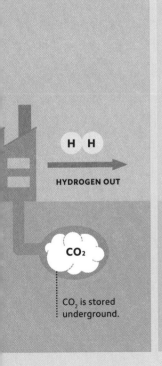

HYDROGEN OUT

CO₂

CO₂ is stored underground.

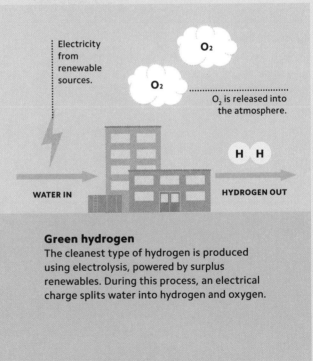

Electricity from renewable sources.

O₂

O₂

O₂ is released into the atmosphere.

WATER IN

HYDROGEN OUT

Green hydrogen
The cleanest type of hydrogen is produced using electrolysis, powered by surplus renewables. During this process, an electrical charge splits water into hydrogen and oxygen.

CUTTING CARBON

Carbon capture is considered a necessary weapon in the fight against climate change, especially when it comes to reducing carbon emissions from "hard to abate" processes—that is, where carbon emissions are difficult to avoid—such as during cement and steel production. All over the world, forests, oceans, peatlands, and other environments naturally capture carbon, and carbon-capture technologies

Capturing carbon at source
The most cost-effective method for capturing carbon is at the point of production, such as in a coal-fired power station or cement kiln.

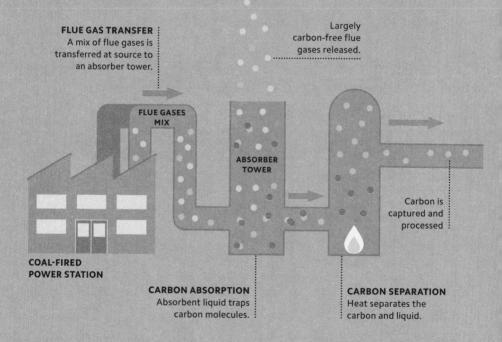

FLUE GAS TRANSFER
A mix of flue gases is transferred at source to an absorber tower.

Largely carbon-free flue gases released.

FLUE GASES MIX

ABSORBER TOWER

Carbon is captured and processed

COAL-FIRED POWER STATION

CARBON ABSORPTION
Absorbent liquid traps carbon molecules.

CARBON SEPARATION
Heat separates the carbon and liquid.

have also been developed. These technologies trap carbon dioxide gas, either at the point of production, such as at a factory, or by drawing it down from the atmosphere. Once carbon has been captured, it is compressed, transported, and usually injected deep beneath the ground for permanent storage. Alternatively, carbon can be recycled for use in various practical applications (see below).

Utilizing carbon

Investment in new technologies continues to open up commercial applications for captured carbon. Some applications do little to reduce emissions. Others, such as injecting captured carbon into concrete, lock carbon up.

"Carbon capture and storage is one of the standout technologies that exists today."

Jade Hameister, Global CCS Institute

CO_2

DIRECT USE

CARBONATED DRINKS

TRANSFORMED CHEMICALLY OR BIOLOGICALLY BEFORE USE

SYNTHETIC FUELS

FERTILIZERS

MANUFACTURING OF MATERIALS SUCH AS CONCRETE OR POLYMERS

GIANT BATTERIES

As intermittent renewables such as wind and solar are added to power grids, there is a need to support them with grid-scale energy storage. This refers to technologies that can store large quantities of energy at times of low demand to release when demand is high, such as storing energy produced by solar farms in the middle of the day for distribution in the evening. Currently, most grid-scale energy storage is achieved through pumped storage hydropower (see below and p.134). However, grid-scale batteries are expected to play a role in supporting grids in the future.

GRID

STORAGE

GRID-SCALE BATTERY
Electricity is turned into chemical energy, which can be stored in grid-scale batteries.

HYDROPOWER
Water pumped uphill to a reservoir is released to make electricity when needed.

GREEN HYDROGEN
Green hydrogen (see p.129) can be stored as a liquid or gas in high-pressure tanks.

NOVEL ENERGY STORAGE
New storage technologies such as compressed air could help balance supply and demand.

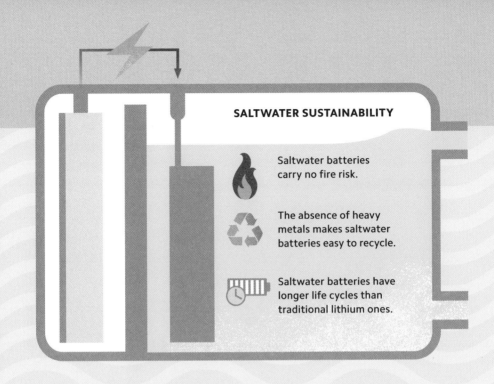

SALTWATER SUSTAINABILITY

Saltwater batteries carry no fire risk.

The absence of heavy metals makes saltwater batteries easy to recycle.

Saltwater batteries have longer life cycles than traditional lithium ones.

BATTERIES THAT DON'T COST THE EARTH

Much more battery capacity will be required to support the phaseout of fossil fuels from the energy sector. Lithium-ion batteries are the most common energy-dense batteries, but they require rare earth metals. Researchers are hoping to create batteries from different materials that perform as well as, or better than, lithium-ion batteries. Among the alternatives are saltwater batteries (see above), which primarily conduct electricity with sodium ions and have the potential to be safe, sustainable, and recyclable. Other options include sodium-ion batteries, solid-state batteries (which have solid electrolytes), and molten-salt batteries (which have salt electrolytes).

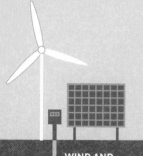

A sand "battery"
Energy can be stored in the form of heat. For example, excess electricity can be used to heat sand stored in insulated silos. The sand remains hot, storing energy for weeks or months at a time.

**WIND AND
SOLAR ENERGY**

HEAT TRANSFER
Hot water heats buildings.

WATER TANK

HOT AIR

HOT SAND
Excess electricity is used to power a heater, raising the temperature of the sand, or sandlike material, to 1,112°F (600°C).

WATER PIPE

HEATING SYSTEM
Air heated by the sand in turn heats water.

BEYOND CHEMICAL BATTERIES

Physical "batteries" offer an alternative to chemical batteries, storing energy in nonchemical form. One example is pumped storage hydropower, using electrical energy to pump water into a raised reservoir and releasing it when power is needed. Approaches that are less reliant on geography are also being explored. These include harnessing gravity by lifting and then lowering heavy blocks; compressing air in underground caverns; heating sand (see above); or accelerating flywheels to extreme speeds.

HARVESTING THE SUN

Solar power is one of the cheapest sources of electricity—their installation is straightforward and relatively little maintenance is required. As a result, there is enthusiasm for spreading panels beyond solar farms. Roof-mounted solar is popular for homes and businesses. Panels can also be fitted to vehicles, such as trains and boats, and infrastructure, such as bus stops and road surfaces. Placing panels on elevated surfaces is preferable, as ground-based panels are easily damaged and get less sunlight.

KITE GENERATORS

Wind power—clean, renewable, and cheaper by the year—constitutes a rapidly growing proportion of the energy sector. Traditional wind turbines could soon be complemented by alternative types of turbines, such as airborne wind turbines—kitelike devices that can be flown at high altitudes where the wind blows hardest. Airborne turbines offer more flexibility than turbines with towers. They can be anchored inland or on barges out at sea, lowered or raised to increase rotation speeds, and deployed in regions prone to hurricanes that are unsuitable for ground turbines.

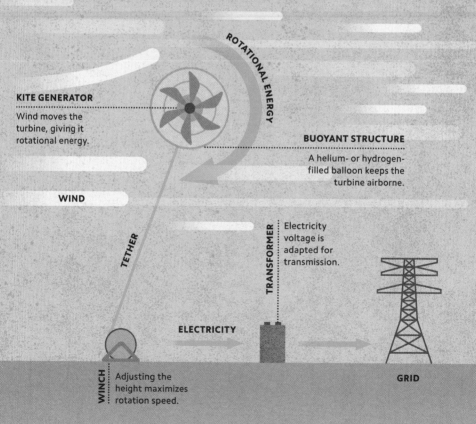

ROTATIONAL ENERGY

KITE GENERATOR

Wind moves the turbine, giving it rotational energy.

BUOYANT STRUCTURE

A helium- or hydrogen-filled balloon keeps the turbine airborne.

WIND

TETHER

TRANSFORMER

Electricity voltage is adapted for transmission.

ELECTRICITY

WINCH

Adjusting the height maximizes rotation speed.

GRID

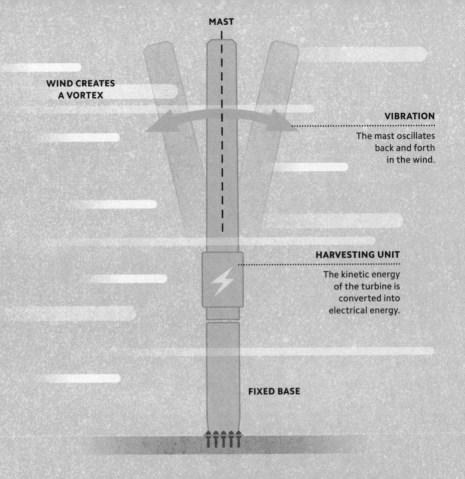

MAST

**WIND CREATES
A VORTEX**

VIBRATION

The mast oscillates
back and forth
in the wind.

HARVESTING UNIT

The kinetic energy
of the turbine is
converted into
electrical energy.

FIXED BASE

GOOD VIBRATIONS

Another alternative to traditional wind turbines, "bladeless"
turbines are relatively small—6–9 ft (2–3 m) high—cylindrical
structures that vibrate back and forth as the wind passes around
them, creating swirling patterns of pressure (vortices). It is
this movement that generates electricity. Bladeless turbines
address some of the common objections to wind turbines. They
are virtually silent to the human ear, do not obstruct views of the
landscape, and do not disturb radar systems or migrating birds.

SCALED-DOWN REACTORS

Nuclear power produces zero carbon and is generally considered a green technology with a role to play in the future of energy. However, new nuclear power stations are among the world's most complicated and costly infrastructure projects. Scaled-down reactors—called small modular reactors (SMRs)—are designed for relative convenience. A fraction of the size of full-size nuclear reactors, SMRs are assembled on site from factory-made components, enabling economies of scale. More than 80 designs are currently under development worldwide.

MICROREACTOR

At 1–20 megawatts (MW), microreactors are small enough to be moved by truck to provide power for remote settlements and military bases.

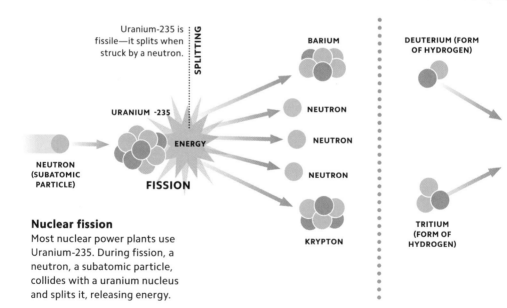

Uranium-235 is fissile—it splits when struck by a neutron.

SPLITTING

URANIUM -235

ENERGY

FISSION

NEUTRON (SUBATOMIC PARTICLE)

BARIUM

NEUTRON

NEUTRON

NEUTRON

KRYPTON

DEUTERIUM (FORM OF HYDROGEN)

TRITIUM (FORM OF HYDROGEN)

Nuclear fission
Most nuclear power plants use Uranium-235. During fission, a neutron, a subatomic particle, collides with a uranium nucleus and splits it, releasing energy.

LARGE-SCALE REACTOR

At 300–1,000+ MW, traditional reactors can provide colossal amounts of power, but planning, constructing, operating, and decommissioning them is a major undertaking.

SMALL MODULAR REACTOR (SMR)

At 20–300 MW, SMRs could be assembled in locations unsuitable for full-size reactors, for example, where access to water is limited.

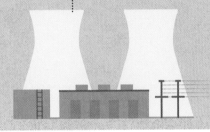

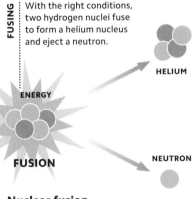

FUSING

With the right conditions, two hydrogen nuclei fuse to form a helium nucleus and eject a neutron.

HELIUM

ENERGY

FUSION

NEUTRON

Nuclear fusion
Nuclear fusion research is focused primarily on the deuterium-tritium reaction. The nuclei of two kinds of hydrogen combine to form helium, releasing energy and a neutron.

ENERGY OF THE FUTURE?

Nuclear power today harnesses nuclear fission—breaking heavier nuclei into lighter nuclei to release energy. A long-held hope is that the opposite reaction, nuclear fusion, which "powers" the stars, could be used to produce unlimited clean energy with no dangerous waste. Fusion is very hard to sustain, requiring extreme pressures and temperatures. Decades of research has seen slow progress, and fusion power is often said to be "always 20 years away".

THE BU
ENVIRO

ILT
NMENT

The built environment refers to the artificially made structures that enable human activity to flourish. It includes buildings, roads, bridges, and parks, and the infrastructure to deliver utilities. Creating and maintaining the built environment for growing populations consumes a huge amount of natural resources. To avoid irreversible environmental damage, more sustainable materials and techniques are being invented, from low-carbon concrete and 3D-printed homes to technology that optimizes our use of services, such as health care or power. Pioneering engineering is also making new kinds of structures possible, for instance, using extraterrestrial regolith to develop built environments on the moon and Mars.

CONNECTED COMMUNITIES

A smart city is an urban environment in which digital technologies collect and analyze data and then use the findings to manage operations and services. This is especially helpful for finding optimal ways to deploy resources, such as electricity, water, road space, and waste collectors, ensuring these respond in real time to changing local demands. It is hoped that smart cities will not only be more environmentally friendly and economically efficient but also provide a better quality of life for their citizens.

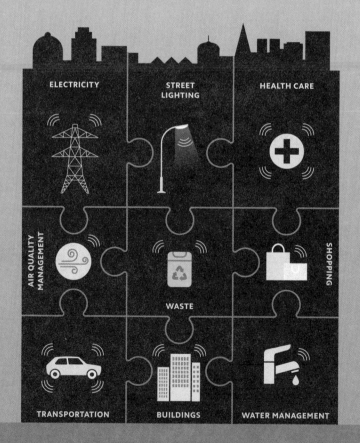

ELECTRICITY

STREET LIGHTING

HEALTH CARE

AIR QUALITY MANAGEMENT

SHOPPING

WASTE

TRANSPORTATION

BUILDINGS

WATER MANAGEMENT

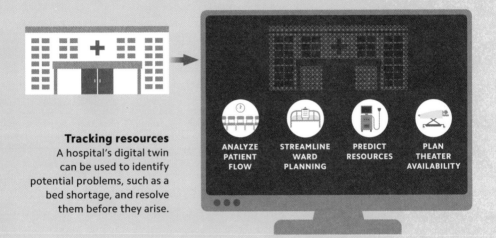

Tracking resources
A hospital's digital twin can be used to identify potential problems, such as a bed shortage, and resolve them before they arise.

ANALYZE PATIENT FLOW

STREAMLINE WARD PLANNING

PREDICT RESOURCES

PLAN THEATER AVAILABILITY

VIRTUAL BUILDINGS

Detailed digital models of physical objects or systems are known as digital twins. They can be created to represent power stations, hospitals, airports, military bases, or even entire cities. These models can then be constantly updated with real-time data. For example, a digital twin of a hospital could be used to monitor and manage patients alongside every category of resource needed to care for them, from paper towels to Accident and Emergency doctors. This can help identify and resolve potentially life-threatening inefficiencies.

ECO HOMES

Greener, cleaner energy usage involves not only phasing out fossil fuels but also using energy far more efficiently. Buildings account for 40 percent of our global energy consumption and are therefore a major focus in the drive to cut energy usage. "Zero-energy" buildings are usually defined as those with net-zero energy consumption. This means that the amount of energy they use over a year is equal to the renewable energy they generate. These buildings are fully insulated to minimize heat loss and have integrated energy-generating devices, such as rooftop solar panels.

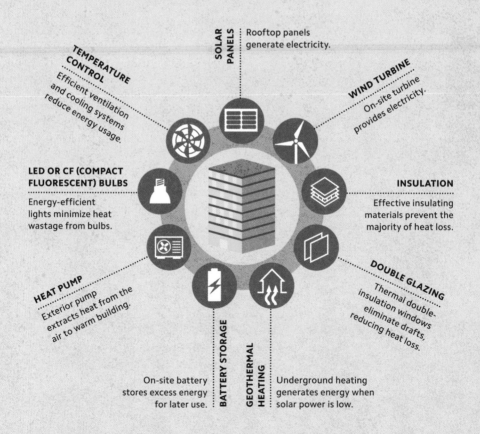

SOLAR PANELS
Rooftop panels generate electricity.

TEMPERATURE CONTROL
Efficient ventilation and cooling systems reduce energy usage.

WIND TURBINE
On-site turbine provides electricity.

LED OR CF (COMPACT FLUORESCENT) BULBS
Energy-efficient lights minimize heat wastage from bulbs.

INSULATION
Effective insulating materials prevent the majority of heat loss.

HEAT PUMP
Exterior pump extracts heat from the air to warm building.

DOUBLE GLAZING
Thermal double-insulation windows eliminate drafts, reducing heat loss.

BATTERY STORAGE
On-site battery stores excess energy for later use.

GEOTHERMAL HEATING
Underground heating generates energy when solar power is low.

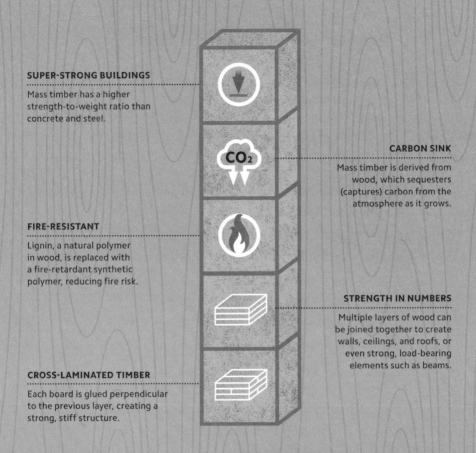

SUPER-STRONG BUILDINGS

Mass timber has a higher strength-to-weight ratio than concrete and steel.

CARBON SINK

Mass timber is derived from wood, which sequesters (captures) carbon from the atmosphere as it grows.

FIRE-RESISTANT

Lignin, a natural polymer in wood, is replaced with a fire-retardant synthetic polymer, reducing fire risk.

STRENGTH IN NUMBERS

Multiple layers of wood can be joined together to create walls, ceilings, and roofs, or even strong, load-bearing elements such as beams.

CROSS-LAMINATED TIMBER

Each board is glued perpendicular to the previous layer, creating a strong, stiff structure.

WOODEN SKYSCRAPERS

People have built with wood for millennia, but advances in engineering mean that wooden buildings can now reach skyscraper heights. Structures may use wood alone or be constructed in combination with concrete and steel. Wooden tower blocks are made possible with mass timber (engineered wood). These wood-derived materials are designed to have certain properties—for example, they can match the strength of concrete and are lighter than both concrete and steel. Cross-laminated timber, made by gluing layers of wood at right angles to each other, is especially strong. Mass timber is renewable and less energy-intensive to produce than conventional building materials.

"GREEN" CONCRETE

After water, concrete is the world's second most used substance. Unfortunately, its production has a heavy environmental impact. Portland cement, concrete's key binding ingredient, is a main contributor of harmful carbon dioxide due to the high temperatures and chemical reactions required to produce it. Efforts to make low-carbon concrete focus largely on finding alternatives to Portland cement. New technologies include binding concrete by injecting captured carbon into the concrete mix (see below). Replacing sand as an aggregate with resources such as plastic waste or concrete rubble also makes concrete greener.

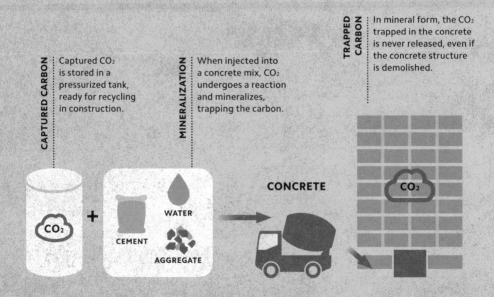

CAPTURED CARBON
Captured CO_2 is stored in a pressurized tank, ready for recycling in construction.

MINERALIZATION
When injected into a concrete mix, CO_2 undergoes a reaction and mineralizes, trapping the carbon.

TRAPPED CARBON
In mineral form, the CO_2 trapped in the concrete is never released, even if the concrete structure is demolished.

CONCRETE

CO₂

WATER

CEMENT

AGGREGATE

CO_2

A carbon negative process
Adding captured CO_2 to concrete means that less cement is needed. The process makes it possible for concrete to be carbon negative— locking away more carbon than it emits.

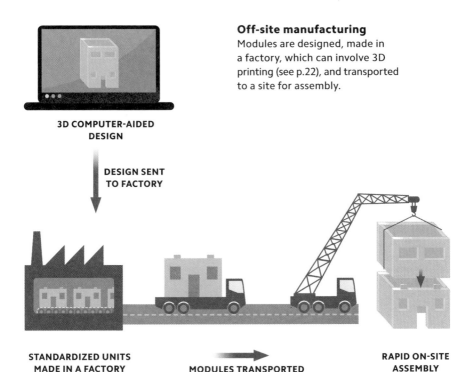

Off-site manufacturing
Modules are designed, made in a factory, which can involve 3D printing (see p.22), and transported to a site for assembly.

3D COMPUTER-AIDED DESIGN

DESIGN SENT TO FACTORY

STANDARDIZED UNITS MADE IN A FACTORY

MODULES TRANSPORTED

RAPID ON-SITE ASSEMBLY

RAPID ASSEMBLY

Modular buildings are made by manufacturing repeating sections— or modules—off-site, then transporting these to a building site for installation. This type of construction has existed in various forms since at least the 1830s but is widely seen as important to the future of construction. It involves minimal waste and environmental disruption, improves efficiency significantly (because multiple parts can be manufactured simultaneously), and provides flexibility. Construction can also take place in environments where conventional methods are unfeasible, for example, at the Halley Research Station in Antarctica.

PRINTED HOMES

With a large enough 3D printer, full-size buildings can be constructed via additive manufacturing (see p.22). Following a digital blueprint, an on-site 3D printer extrudes a pastelike mixture, usually concrete, in successive layers to form the structure. Plumbing, wiring, and components such as windows and doors are added later. Building in this way is quick, uses resources efficiently, and allows more creative freedom, at a fraction of the cost of conventional construction.

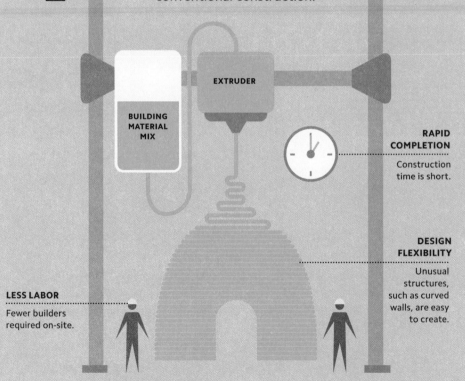

EXTRUDER

BUILDING MATERIAL MIX

RAPID COMPLETION
Construction time is short.

DESIGN FLEXIBILITY
Unusual structures, such as curved walls, are easy to create.

LESS LABOR
Fewer builders required on-site.

FLOOD MITIGATION

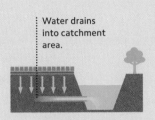

Water drains into catchment area.

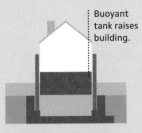

Buoyant tank raises building.

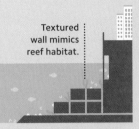

Textured wall mimics reef habitat.

PERMEABLE SURFACES
Gaps between tiles allow rainwater to soak through.

FLOATING BUILDINGS
Buoyant foundations elevate buildings during floods.

ECO-FRIENDLY SEA WALLS
These protect against storms and support biodiversity.

CLIMATE PROOFING

The frequency and intensity of extreme weather events are increasing. In addition to choosing not to build in areas at risk from rising sea levels, we can adapt future infrastructure to be more climate-resilient. For instance, railroad tracks could be made from heat-resistant metals less prone to expanding and buckling in very hot weather, and buildings can be covered with plants to create shade or even located underground.

HEAT MANAGEMENT

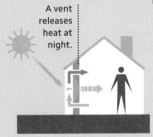

A vent releases heat at night.

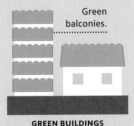

Green balconies.

TROMBE WALL
A glass panel absorbs and stores sunlight during hot weather.

GREEN BUILDINGS
Plant-covered buildings absorb pollutants and create shade.

SOLAR-REFLECTIVE SURFACES
A white roof reflects sunlight, keeping temperatures inside low.

MOVING PARTS

Large structures that have the ability to adapt in response to changes to their environment are known as active structures. Spacecraft tend to be active structures to enable them to adapt to extreme conditions during their missions. The International Space Station, for instance, can orient and retract attached 115 ft (35 m) long solar arrays. Hypothetical giant structures, such as space elevators or towers reaching from the equator into space, would also be active structures.

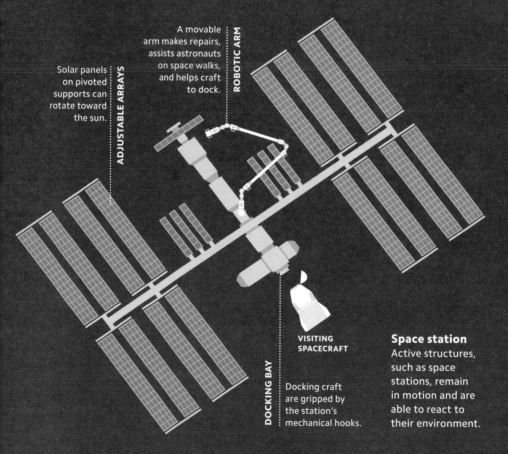

ADJUSTABLE ARRAYS
Solar panels on pivoted supports can rotate toward the sun.

ROBOTIC ARM
A movable arm makes repairs, assists astronauts on space walks, and helps craft to dock.

VISITING SPACECRAFT

DOCKING BAY
Docking craft are gripped by the station's mechanical hooks.

Space station
Active structures, such as space stations, remain in motion and are able to react to their environment.

POD LIFE

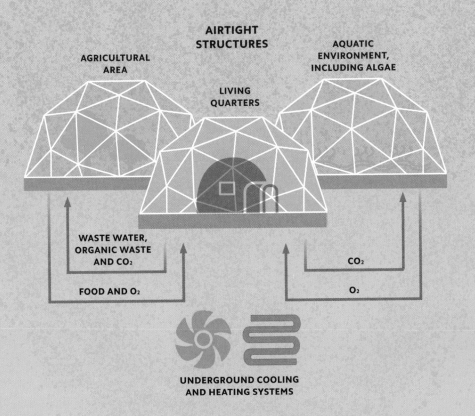

AIRTIGHT STRUCTURES

AGRICULTURAL AREA

LIVING QUARTERS

AQUATIC ENVIRONMENT, INCLUDING ALGAE

WASTE WATER, ORGANIC WASTE AND CO_2

CO_2

FOOD AND O_2

O_2

UNDERGROUND COOLING AND HEATING SYSTEMS

Closed ecological systems are living communities that can exist in total isolation—that is, without exchanging matter, such as oxygen, with the external environment. Earth could be seen as one such example, but the term usually refers to small, artificial ecosystems. In a closed system, waste from one species is used by another, for instance, as food. While there is no immediate need for such systems, research is ongoing, with a view to building bases on the moon and Mars.

BUILDING WITH SPACE DUST

If humans are to live for extended periods on the moon, Mars, or other astronomical bodies, they will not be able to rely indefinitely on resources sent from Earth. This means finding ways to survive—including constructing habitats—using the materials available in situ. Space agencies are preparing for future missions by exploring ways to use regolith (the material found on the surface of a moon or planet) as the basis for construction.

Working with space materials

Sintering, binding, and cold welding can all be used to build with regolith.

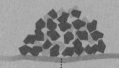

RAW MATERIAL

The moon and Mars are covered with a loose mixture of dust and rock —or regolith.

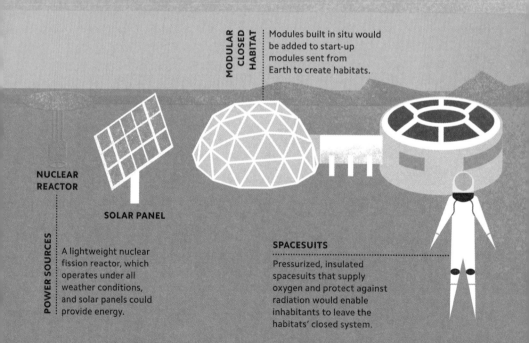

MODULAR CLOSED HABITAT

Modules built in situ would be added to start-up modules sent from Earth to create habitats.

NUCLEAR REACTOR

SOLAR PANEL

POWER SOURCES

A lightweight nuclear fission reactor, which operates under all weather conditions, and solar panels could provide energy.

SPACESUITS

Pressurized, insulated spacesuits that supply oxygen and protect against radiation would enable inhabitants to leave the habitats' closed system.

SINTERING

Concentrated sunlight can sinter regolith—that is, heat it into a solid.

BINDING

Materials available on site, such as melted sulfur or phosphoric acid, can bind regolith.

BACKUP BINDER

As an emergency backup, human blood could be used as a binder.

WELDING

Regolith can be "cold" welded, using pressure between smooth surfaces to form atomic bonds.

REGOLITH BUILDING

Local materials can be used to help create modular habitats on the moon and Mars.

LIFE ON MARS

Colonizing space has long been a staple of science fiction. However, without terraforming (see pp.154–155), living on an astronomical body such as Mars would require artificial habitats with complex life-support systems to protect against the deadly atmosphere, low pressure, solar radiation, and extreme cold. Scientists are currently exploring how to achieve this in terms of in-situ resources (see above) and other aspects of extraterrestrial living.

ROVERS

Robotic vehicles could carry out practical tasks on the surface, such as collecting regolith.

EARTH 2.0

Terraforming (which literally means "Earth-shaping") is the deliberate transformation of an astronomical body to make it more Earth-like. This is usually in the speculative context of making it habitable by humans without the use of artificial, closed ecosystems (see p.151) and spacesuits. Terraforming Mars—the planet with the most potential to

MARS TODAY

Plants or algae produce oxygen.

Heat is trapped.

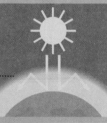

USE GREENHOUSE EFFECT
Raising the concentration of atmospheric carbon on Mars could trap heat from the sun on the surface.

INCREASE OXYGEN
To make the atmosphere on Mars breathable, the oxygen content would need to be raised.

CAUSE A DIRECT IMPACT
Steering an asteroid into Mars could provide energy, slightly raising temperatures and perhaps creating a lake.

An inhospitable planet
A surface temperature as low as −225°F (−153°C) and a thin atmosphere make Mars inhospitable to life. Terraforming could involve introducing plant life and manipulating the surface and atmosphere to provide heat and water.

be hospitable to humans—would require raising its surface temperature, importing water, and creating a breathable and protective atmosphere. Although the terraforming of Mars is likely to be beyond the limits of existing technology, it has been seriously examined by scientists as a very long-term possibility for securing humanity's future.

Less heat loss.

MANIPULATE REFLECTION
Darkening the surface of Mars reduces the reflection of sunlight so more heat is absorbed.

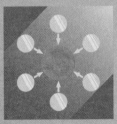

CONCENTRATE THE SUN
Orbiting mirrors could focus sunlight, boosting heat and light for use in solar power.

CREATE A MAGNETIC FIELD
Magnetic fields protect planets from cosmic rays, and there are several suggestions as to how to create one for Mars.

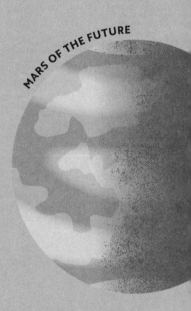

MARS OF THE FUTURE

INDEX

Page numbers in **bold** refer to main entries.

ACKNOWLEDGMENTS

DK would like to thank the following for their help with this book: Debra Wolter for proofreading; Helen Peters for the index; Senior DTP Designer Harish Aggarwal; Senior Jackets Coordinator Priyanka Sharma Saddi.

All images © Dorling Kindersley
For further information see:
www.dkimages.com

SIMPLY EXPLAINED

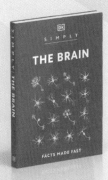

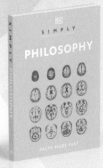

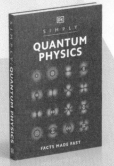